Toys and Communication

Luísa Magalhães · Jeffrey Goldstein
Editors

Toys and Communication

palgrave
macmillan

Editors
Luísa Magalhães
Faculty of Philosophy and Social Scienes
Catholic University of Portugal
Braga, Portugal

Jeffrey Goldstein
Institute for Cultural Inquiry
Utrecht University
Utrecht, The Netherlands

ISBN 978-1-137-59135-7 ISBN 978-1-137-59136-4 (eBook)
https://doi.org/10.1057/978-1-137-59136-4

Library of Congress Control Number: 2017939098

Cover illustration: culliganphoto/Alamy Stock Photo

Printed on acid-free paper

This Palgrave Macmillan imprint is published by Springer Nature
The registered company is Macmillan Publishers Ltd.
The registered company address is: The Campus, 4 Crinan Street, London, N1 9XW, United Kingdom

Preface

Toys and Communication has its origins in an International Toy Research Association conference on this theme in July 2014 in Braga, Portugal: http://itratoyresearch.org. We have asked authors of some of the most interesting papers to expand them into chapters for this book. The contributing authors include established play and toy researchers, among them Cleo Gougoulis, Gilles Brougère, Steven Kline, Michel Manson, David Myers, and Jean-Pierre Rossie, as well as promising younger scholars. All chapters were prepared exclusively for this book and have not been previously published. Several of the papers that serve as the basis for these chapters have won research awards. Among the chapters in *Toys and Communication* are explorations of images of toys in art (Vaz and Manson), toys and cultural transmission in Morocco (Rossie) and India (Patil), and within the family (Wade), toys and language and communication skills (Gummer; Maggio, Phillips and Madix). Toys during and about the Second World War are the subjects of Gougoulis' description of children's toy play in Greece during the German Occupation and of Seriff's chapter on toys as propaganda. In the chapters by Kline and by Vaz and Manson the representation of toys in advertising and in art is considered, while Carla Ferreira and Luisa Agante present their experiment on toy premiums and their influence on healthy eating habits. Spaces designed for play and the design of playthings themselves, are the subjects of chapters by Magalhães, van Leeuwen and Gielen, and Leclerc.

Several chapters in *Toys and Communication* stress the child's use of toys to 'become' more adult. Brougère, Gummer, Wade, Gougoulis, and Magalhães all refer to toys and situations that encourage the acquisition of adult roles and behaviors.

Braga, Portugal — Luísa Magalhães
Utrecht, The Netherlands — Jeffrey Goldstein

Acknowledgements

The editors would like to thank:

Catarina Leite for her efficient and priceless help throughout the editing process
Carolyn Sheils for her advice on toys in speech therapy
Heloise Harding for her professionalism and dedication.

Contents

About the Editors

Luísa Magalhães, Ph.D. is a researcher in Communication Sciences, at CEFH, Center for Philosophical and Humanistic Studies, Catholic University of Portugal and CECS Communication and Society Research Center, University of Minho. She is an ITRA Board member and was chair of the 7th ITRA World Congress, 2014.

Jeffrey Goldstein is chairman of the National Toy Council (London. http://www.btha.co.uk/value_of_play/toy_council.php) and chairman of the Experts Group of PEGI, the European video games rating board (www.pegi.info). He is co-founder with Brian Sutton-Smith and Jorn Steenhold of the International Toy Research Association (www.itratoyresearch.org).

List of Figures

List of Tables

PART I

Toys and Communication. Preliminary Issues

CHAPTER 1

Toys and Communication: An Introduction

Jeffrey Goldstein

Preliminary Issues

Toys and their meanings in play, and sometimes in earnest, can be seen as aspects of communication and the transmission of culture (Goldstein 2011). Steve Kline retraces his intellectual journey through children's culture and toys, indicating the ways in which they can be regarded as 'media.' Reflecting on 30 years as a media researcher, Kline traces the marketing of toys and games in order to re-situate debates about children's toy play within the context of a global market culture. Kline interrogates the conflicting value discourses surrounding creativity and imagination in the age of 'digital play media.'

Gilles Brougère asks whether toys are educational. What relationship can there be between an object and the act of learning? These questions often surround the discourse regarding toys and are caught in a tension that shows why such questions are poorly formulated, even meaningless. Indeed, the contradictory answers to these questions reflect two ideas (which can sometimes happily meld together): all toys are educational; certain toys are educational.

J. Goldstein (✉)
Institute for Cultural Inquiry, University of Utrecht,
Utrecht, The Netherlands
e-mail: J.Goldstein@planet.nl

L. Magalhães and J. Goldstein (eds.), *Toys and Communication*,
https://doi.org/10.1057/978-1-137-59136-4_1

A toy is mainly an object addressed to children. It must persuade children or their caregivers. To analyze this kind of communication, we can use the notion of rhetoric—that is, what the toy means to its recipient, and how the object captures the consumer and/or the user, a child or adult on behalf of the child. The main way today is to make the object fun, mainly in relationship to children's mass culture; that is, a rhetoric of fun. Another way is to connect children with education or development. Toys for children are seen as educational tools for children. But if a rhetoric of education has become less central and important for the toys, it still perpetuates this mainly for younger children who are more dependent on the adult's choice. Brougère argues that the rhetoric of education is more a discourse, an image, than a reality.

David Myers (this volume) considers the ways in which toys and games occupy different spaces. Over the past quarter-century, digital media have had a large impact on toys and toy production. The USA National Toy Hall of Fame, for instance, now includes the Atari 2600 and Nintendo Game Boy alongside alphabet blocks and jacks as 'classic icons of play' (http://www.toyhalloffame.org/). Myers considers the ways in which toys and games are dissimilar. Early successful digital adaptations of popular toys (recasting the teddy bear as Teddy Ruxpin, for example) have yet to supplant their non-digital predecessors as overwhelmingly as *World of Warcraft* (2004) and similar Massive Multiplayer Online games have supplanted *Dungeons and Dragons* (1977).

There are several dimensions without need of reference to digital media along which toys and games might be distinguished. These include toys being associated with younger players (relative to game players), the physicality of the toy (relative to a digital game), and the rule-dependent nature of the game (relative to the toy). Myers suggests that these and related distinctions are rooted in the unique semiotic properties of the toy and the game. In most circumstances, in fact, digital media impose a fundamental dissonance during toy play: a context of control.

Language Development

Amanda Gummer (this volume) examines the roles of toys in the development of children's language and communication. To offer a concrete example, consider playing with blocks. Playing with blocks promotes language development, spatial skills, and basic mathematics concepts. These conclusions are based on studies of children from 1½ to 6 years of age by Christakis, Zimmerman and Garrison (2007).

There are two main types of play with blocks. In *free play*, which is unstructured block play, children use blocks to build designs of their choice. In *structured block play* children attempt to make a particular structure from a model or plan. These two types of play involve different processes: the former relies on children's imagination and ability to produce complex relations without prompting, while the latter calls upon the ability to analyze a spatial representation to create a predefined model. Free play with blocks and construction toys requires imagination, creativity, and planning. The basic form of building blocks allows constructions to be as complex as the imagination of the child using them. Structured block play, such as copying a model, requires careful observation, counting, and spatial relations. Both types of block and construction play contribute to early mathematics skills and language development.

In one study, children who played with plastic building bricks had significantly better language skills six months later, compared to a control group. In the study (Christakis et al. 2007), sets of molded plastic building bricks were distributed to a random group of families with children aged 1½–2½ years who were registered at a pediatrics clinic. The parents also received newsletters with 'blocktivities' suggesting things that they and their child could do with the blocks (sort blocks by color, see how big a stack they could make, etc.). Data were gathered from 140 families. Most of the children who received blocks reported playing with them. Six months later, middle-and low-income children who received blocks had significantly higher language scores. The researchers conclude that playing with blocks can lead to improved language development in children from middle-and low-income families.

Why does block play promote language? As children manipulate objects, they begin to understand more about their qualities and relations. Older children begin to make up stories or scripts for these objects, which underlie further understanding of them. One theory that may explain associations between early exposure and subsequent cognitive and linguistic outcomes is based on the development of mental schemas that Vygotsky referred to as 'scaffolding.' (See Gummer, this volume.) Mental schemes are internal models of the world that a child uses to understand and master his or her environment. They are the precursors of thought and language. *'Through play, that is, unstructured manipulation of objects, the child begins to develop a mental picture of and cognitive categories about the objects around him or her. These mental schemes underlie an understanding of object permanence, the development of memory, and the roots of impulse control and language'* (Christakis et al. 2007). An important leap in understanding occurs when the child learns to substitute and combine mental categories internally. For example, a bath sponge used as a boat.

Tina Bruce (2008) describes a study in which childcare providers were offered blocks and training in their use. The aim was to explore how block play can support sustained learning and development for children aged 2 to 5. Two groups were involved in the project. 48 pre-schools and day nurseries were offered in-depth training. They received a school set of Community Playthings unit blocks and a camera to record their projects. The 48 child-minders received a one-day training course in the use of these objects. Practitioners found that, through block play, children became better able to regulate disagreements, boys and girls were more likely to play together, and children developed an awareness of risk and how to manage it. The blocks supported a richer diversity of play, holding children's interest over longer periods of time. Because block play transcends differences in age or background, older and younger children work cooperatively. Even if children do not speak the same language, they understand each other as they create with blocks. They develop ideas together. Children converse as they play, suggesting how play assists communication.

According to Whitebread and Basilio (2014), there are three characteristics of toys that promote early development: (1) Contingency: something clearly changes as a result of the child's actions; (2) Cognitive challenge: there is a clear goal but it is challenging to achieve; (3) Scaffolding and control: adults can adjust the level of difficulty of the goal while the child maintains control of the actions.

Different toys elicit different speech. Children's play and language tend to follow the themes suggested by their toys (Vukelich et al. 2013). Functionally explicit or realistic toys, such as a doctor's kit or a truck, stimulate related play themes and language. Toys offering a fantasy theme, such as dolls or dinosaurs, evoke fantastic play and decontextualized language(Pellegrini and Jones 1994). For example, children use more varied functions of language when playing with housekeeping toys, such as pots, pans, stoves, and refrigerators, than when playing with art materials, such as paper, pencils/crayons, scissors, glue, and water/sand tables.

Toys also serve as prompts that stimulate parent-infant communication, which is crucial to language development(Roseberry et al. 2014). Experiences with spatial toys such as blocks, puzzles, and shape sorters, and the spatial words and gestures they evoke from adults, have a significant influence on the early development of spatial skills, which are important for success in science, technology, engineering, and mathematics (STEM) (Verdine et al. 2014).

Amanda Gummer describes the chronological development of communication in play and refers to specific toys and activities that were observed to facilitate the process. She compares the differences in the use of toys as communication tools by children aged 4–6 years with those aged 10–11 years. Observations are made of children's use of toys to communicate with both peers and adults, and the different types of toys chosen by the children for a range of communications (e.g. intentional and non-intentional). The emergence of a sense of identity is believed to be the critical developmental factor to explain the differences observed between the two age groups. Gummer further considers other factors such as gender differences, ethnic origin, and socioeconomic factors.

By using toys with forms and aesthetics that are strongly influenced by typographic communication, there is potential for hybrid symbols to influence visual recognition of letters for literacy and language during toddler and preschool years. The type-as-toy concept of Maggio, Phillips and Madix has potential in a wide range of toys bound by the same design methodology—the first being a hanging mobile selected for its inherent benefits in early cognitive development. Combined with caregiver interaction and joint attention, Maggio et al. investigated whether children in preschool can identify typographic symbols arranged into zoomorphic compositions in the context of a toy.

Music and Cognition

Toys that enable listening to, or playing, music involve multiple senses and multiple regions of the brain. Music is processed by different areas of the brain working closely together to make sense of things such as melody, harmony, and rhythm. The ability to recognize patterns is necessary to appreciate music; it is also a key component of intelligence. So we should expect a relationship between musical comprehension and some forms of intelligence, including spatial reasoning, memory, language, math comprehension, and creativity (Weinberger 1998).

'Brain plasticity' is the term for the capacity of the human brain to alter in response to the environment. Hearing music affects brain plasticity and the way a child's brain develops. Exposure to music prenatally appears to help postnatal coordination and physical skills. In one study, babies in utero were exposed to recordings of different musical components that increased in complexity during the pregnancy. In total, each baby was exposed to between 50 and 90 h of music. The babies exposed to music seemed to make faster progress in some areas compared to babies who had not been exposed to music. Their pre-speech became evident sooner, and there were noticeable differences in hand-eye coordination, visual tracking, facial mirroring, general motor coordination, and the ability to hold a bottle with both hands (Lafuente et al. 1997).

Children from preschoolers through adolescents who receive from six months to two years of piano lessons show improved cognitive skills, spatial skills, and vocabulary compared to control groups that did not receive music lessons (Piro and Ortiz 2009). Music participation, both inside and outside of school, is associated with measures of academic achievement among children and adolescents. In one study, listening to a Mozart piano sonata increased spatial-temporal reasoning (Rauscher et al. 1997). Adult musicians' brains show clear differences from those of non-musicians, particularly in areas relating to listening, language, and the connection between the two sides of the brain (Hyde et al. 2009).

In sum, music, in addition to its mood management function, also affects brain development, language, and cognitive development (see Goldstein 2015).

Toys, Culture, and Communication

Anthropological literature on children's self-made toys tends to focus more on toy construction in rural non-western areas of the world, where toy making is part of a tradition of craftsmanship in self-subsistence economies (Patil, this volume; Rossie, this volume).

Giving toys to children tells us something about the child, the donor, and the relationship between them. Depictions of children in art reveal similar information, as decoded by Vaz and Manson, who analyze their cultural messages. Portraits are strongly subject to the constraints of their sponsors, who saw the opportunity to strengthen their religious and sociopolitical aspirations. The child portrait with toy(s) is examined, focusing on the princely children of the Spanish Golden Age (sixteenth to eighteenth centuries). Musical toys appear traditionally within aristocratic families but undergo a shift to middle-class families in the eighteenth and nineteenth centuries. Other toys have gradually appeared in portraits examined by Vaz and Manson.

Jean-Pierre Rossie highlights Moroccan children's self-made and imported toys as these are used in play activities through which information, feelings, attitudes, and behavior are transmitted to peers and younger children. This horizontal transmission can be contrasted with the better-researched vertical transmission between adults and children as found in socialization, training, and non-formal education. The content of this transmission between Moroccan children refers to the material environment, the techniques, the beliefs, the verbal and non-verbal behavior, and the relations in the families to which they belong. Toys during this transmission process emerge as signs, as non-verbal language, as objects in specific sociocultural contexts, as creative manifestations and as gendered objects. Attention is paid not only to more traditional situations but also to recent changes, such as play and toy-making activities influenced by television programs.

Using the creation and variety of playthings in Banaras, India, Koumudi Patil offers a case study in the community of discourse that develops among the children, parents, and creators of playthings. Here, too, we see how toys both reflect and shape culture, and how change over time comes about. Patil's doctoral dissertation was a study of workshops in Banaras, India, where craftsmen turn out traditional toys. Toys in Patil's work emerge as a means used by local craftsmen to negotiate their identity in the face of a changing global context exemplified by the growth of a tourist market. Banarasi craftsmen incorporate changes by assigning different uses to toys. Counter to dominant perceptions of traditional production methods as conservative, Patil shows how the structure of the traditional toy assembly line leaves scope for change and innovation. Furthermore, Patil adds meticulous analysis and ethnography to examine and expand upon (Dundes' 1993) interpretation of a Banarasi pecking toy.

Focusing on oral historical interviews and archival museum research, Suzanne Seriff takes a close look at the multiple ways in which toys are appropriated, marketed, circulated, and manipulated for ideological, economic, or educational purposes by artists, politicians, and toy companies for purposes beyond their original intention as playthings for children. She examines two 'toys,' a 1930s' anti-semitic board game from Nazi Germany and a work presented as art that uses Lego bricks to form a replica of a concentration camp. Seriff suggests that anti-semitic toys and games are not innocuous but convey powerful messages, that, although they may not incite individuals to act, could at least lend support to those who might victimize Jews.

Based on interviews and toy-making sessions with the elderly in a Day Care Center of a working-class neighborhood in Piraeus, Greece, anthropologist Cleo Gougoulis describes the ingenuity and inventiveness of urban working-class children during World War II. During the German Occupation and the ensuing year of famine that plagued the major Greek urban centers, children of the poor not only had to work to survive but also struggled to find the materials for constructing their toys in times when materials for toy making were as scarce as food. Gougoulis considers the work and the time invested in toy making as a dynamic, complex process uniting work and play, leisure and labor, and promoting children's active participation in the adult world (see also Magalhães, this volume).

Toy premiums, such as those that accompany fast-food meals, may affect young children's food attitudes and consumption. Carla Ferreira and Luísa Agante present their study with school-age children, asking the extent to which toy premiums promote healthy eating habits. What do toy premiums communicate to their recipients? Previous research examined their effects on the consumption of cereals, among children under 12. Ferreira and Agante ask whether older children are similarly affected. The chapter dovetails nicely with that of Mariah Wade on miniature cooking sets and children's attitudes towards food.

Similar to miniature porcelain dining and serving sets of the past, today's toy baking sets, stoves, cooking utensils, aprons, and table linens train the mind and the body to assumed gender-specific chores. Playing at becoming, children mimic a mother's routine of food selection, preparation, display, and consumption. With and through cooking and serving toys, children develop a 'sensorial compass,' as Wade calls it, and an affect that will define their culinary likes and dislikes, their recognition of food, home, and comfort. Equally, these mimetic performances of domesticity reproduce a culture of food and display expectations that become familial traditions. Wade draws on ethnography, oral history, and archival research to explore the triple process whereby female youngsters through play become socialized to foodways, associate those foodways with their '"elected" or "lost" homelands,' and learn to cherish expatriate otherness through food traditions.

Toy Design and Play Spaces

The chapters by Leclerc, van Leeuwen and Gielen, and Magalhães describe ways in which play influences toy design, and design influences play. Leclerc considers the multiple forms of play as inspiration for toy design. Van Leeuwen and Gielen begin with the reversal theory of motivation (Apter 2007) to show how the need for autonomy and competence can be addressed through design.

'KidZania' is a theme park that provides a universe of training and of fun for children to practice and execute adult jobs, and to explore their ability to imitate the world of adults, including their possible market choices. This park consists of a 'city' built to children's scale where children can play at being 'adults,' choosing from more than 60 different professions in replicas of the most representative institutions of a real city: airport, factories, shops, police stations, firefighters, press, TV studio, and stadium, among others. There even exists a special currency that is earned in the various jobs and spent in the city afterwards. The author considers the use of toys within the different professional contexts that are offered by the park. Are children playing in the KidZania theme park? Are they working? Are children engaged in role-play activities or are they really committed to performing a job? And what distinguishes these concepts as far as children are concerned? Magalhães administered a questionnaire to 300 children to examine their understanding of play and their response to the performance of adult jobs, and corresponding handling of currency within the context of a real bank. Data show the willingness of children to return to KidZania Park as an intention to carry on with adult life situations. This confirms Gilles Brougère's formulation about play as the major expression of children's desire to be adult.

Rémi Leclerc offers a philosophical and pedagogical look at design for play, and at how both play and design create and recreate culture. Leclerc says the much-heralded rise of a leisure society has not materialized. Instead, many of us face stress, long working hours, and unhealthy lifestyles, common conditions for most people around the globe, and especially perhaps in the major cities of Hong Kong and China. Also, while the media and the fashion industries have reshaped sports into a spectacle, leisure activities have become branded commercial affairs, thus defeating the purpose of leisure: as they consume leisure-as-product experiences, citizens are alienated from their individual and communal quest for recreation.

Leclerc presents the work developed by Hong Kong Polytechnic University School of Design's undergraduate Industrial and Product

Design students in the Design for Leisure program. It highlights how students embraced the Hong Kong urban context, and proposed a number of projects with quirky interpretations of such urban games as the Situationists' 'Dérive,' the Treasure Hunt, or Eco Warriors' 'Green Bombs.' The project aimed to provide design students with the necessary skills to critically assess how humans have come to separate play from work, and contextualize the social, cultural, emotional and physical importance of outdoor play and exercise in modern society.

Lieselotte van Leeuwen and Mathieu Gielen propose the need to support rebelliousness in children, and explore whether and how design for play could support this. They note that autonomy, resilience, assertiveness, self-efficacy, self-regulation, internalization of moral rules, and creativity all implicitly require non-compliance to some degree. They employ developmental psychology and Apter's Reversal Theory of motivation (Apter 1982, 2007) to inform design. At first sight, design for rebelliousness seems a contradiction in terms since a design 'asking' to break rules would require conforming to the provided design idea. While rebellious characters in children's stories are loved and embraced, the existing incorporation of rebelliousness into active play contexts largely mirrors the existing bias towards 'right' and 'wrong.' Van Leeuwen and Gielen describe principles of toy design that can encourage rebelliousness.

In sum, toys are found to speak to us and through us, communicating messages about their uses, values and meanings.

References

Apter, M. J. (1982). *The experience of motivation: The theory of psychological reversals.* London: Academic Press.

Apter, M. J. (2007). *Reversal theory: The dynamics of motivation, emotion and personality.* Oxford: Oneworld.

Bruce, T., McNair, L., & Siencyn, S. W. (2008). *I made a Unicorn! Open-ended play with blocks and simple materials.* Robertsbridge UK: Community Playthings.

Christakis, D. A., Zimmerman, F. J., & Garrison, M. M. (2007). Effect of block play on language acquisition and attention in toddlers: A pilot randomized controlled trial. *Archives of Pediatric & Adolescent Medicine, 161*(10), 967–971.

Dundes, A. (1993). *Folklore matters.* Knoxville, TN: University of Tennessee Press.

Goldstein, J. H. (2011). Play and technology. In A. Pellegrini (Ed.). *Oxford handbook of the development of play* (pp. 322–340). Oxford University Press. ISBN 13: 978-0195393002.

Goldstein, J. H. (2015). Applied entertainment: Positive uses of entertainment media. In R. Nakatsu, M. Rauterberg, & P. Ciancarini (Eds.). *Handbook of digital games and entertainment technologies.* Springer. doi:10.1007/978-981-4560-42-8_9-1. ISBN 978-981-4560-49-8.

Hyde, Krista L. et al. (2009). Musical training shapes structural brain development. *Journal of Neuroscience,* 29, 3019–3025.

Lafuente, M. J. et al. (1997). Effects of the Firstart method of prenatal stimulation on psychomotor development: The first six months. *Pre- and Perinatal Psychology Journal,* 11, 151.

Pellegrini, A. D., & Jones, I. (1994). Play, toys, and language. In J. H. Goldstein (Ed.). *Toys, play and child development* (pp. 27–45). Cambridge University Press.

Piro, J. M. & Ortiz, C. (2009). The effect of piano lessons on the vocabulary and verbal sequencing skills of primary grade students. *Psychology of Music,* 37, 325–347.

Rauscher, F. H. et al. (1997). Music training causes long-term enhancement of preschool childrens' spatial-temporal reasoning. *Neurological Research,* 19, 2–8.

Roseberry, S., Hirsh-Pasek, K., & Golinkoff, R. M. (2014). Skype me! Socially contingent interactions help toddlers learn language. *Child Development, 85*(3), 956–970.

Sutton-Smith, B. (1997). *Ambiguity of play.* Cambridge MA: Harvard University Press.

Verdine, B. N., Golinkoff, R., Hirsh-Pasek, K., & Newcombe, N. S. (2014). Finding the missing piece: Blocks, puzzles, and shapes fuel school readiness. *Trends in Neuroscience and Education, 3,* 7–13.

Vukelich, C., Christie, J., & Enz, B. J. (2013). *Helping young children learn language and literacy. Birth through kindergarten* (3rd ed.). New York: Pearson.

Weinberger, N. (1998). Brain, behavior, biology, and music: Some research findings and their implications for educational policy. *Arts Education Policy Review, 99,* 28–36.

Whitebread, D., & Basilio, M. (2014). *Thinking with your hands: How can toys for infants and toddlers foster early learning and communication?* Paper presented at 7th World Congress, International Toy Research Association. Braga, Portugal.

Author Biography

Jeffrey Goldstein is chairman of the Experts Group of PEGI, the European video games rating board (www.pegi.info). He is co-founder with Brian Sutton-Smith and Jorn Steenhold of the International Toy Research Association (www.itratoyresearch.org).

CHAPTER 2

The End of Play and the Fate of Digital Play Media: A Historical Perspective on the Marketing of Play Culture

Stephen Kline

Introduction

As Faulkner famously wrote, 'The past is not dead. Actually it is not even past.' As I begin to write this chapter, a legend about an imagined galaxy long long ago is preparing an invasion of global toy stores with *Star Wars* merchandise whose promised sales could amount to 4 billion dollars this Christmas season. Meanwhile, the hacking of a *Hello Kitty* website has released children's private information to the world, which perhaps is why the Campaign for a Commercial Free Childhood (CCFC) advocacy group is awarding their 'TOADY' (Toys Oppressive And Destructive to Young children) for the holiday season's worst toy to *Hello Barbie*, a networked doll that enables Mattel to analyze children's language produced while playing. And the purpose of all this commercial research into children's play culture is the desire to sell more toys. With these developments in mind, I want to argue that the theme of the 2014 International Toy Research Association (ITRA) conference—analyzing toys as communication and language—remains as relevant now as when a small group of toy

S. Kline (✉)
Simon Fraser University, Vancouver, BC, Canada
e-mail: kline@sfu.ca

L. Magalhães and J. Goldstein (eds.), *Toys and Communication*,
https://doi.org/10.1057/978-1-137-59136-4_2

Fig. 2.1 International toy research association, 1993

researchers (pictured here) were convened by Brian Sutton-Smith, Jeffrey Goldstein, and Jorn-Martin Steenhold to reflect what was happening to play culture in the 1990s (Fig. 2.1).

International Toy Research Association, 1993 First meeting of toy researchers, Utrecht University. From left to right: Rachel Karniol, Stephen Kline, Gisela Wegener-Spohring, Hein Retter, Waltraut Hartmann, Jean-Pierre Rossie, Brian Sutton-Smith, Peter K. Smith, Kathleen Alfano, Gilles Brougère, Birgitta Almqvist, Anthony Pellegrini, Jeffrey Goldstein, Maria Bartels (assistant), Greta Fein, Jorn-Martin Steenhold.

It is said that those who fail to learn from history are destined to repeat it. It is also said that, as one gets older, one's hope for the future fades while history becomes more vivid in the present. In this essay, I argue that our commentary on digital play media in the twenty-first century needs to be informed by the cultural history of toys as communication media. We may not have reached the end of history (Kline 2015), but the marketing strategies promoting playfulness in the digital media environment of today remain a central theme in the critical analysis of play cultures. Indeed, I will argue that the study of children's play culture at the turn of the millennium, as well as the emerging critiques of children's media-saturated lives, can especially benefit from analyzing the market dynamics galvanizing contemporary play culture in the postindustrial market. But before I give this history of modern toy marketing, I hope the reader will indulge this aging historian of play media for reviewing the intellectual history underwriting ITRA's decision to refocus our play research on the analysis of language and communication.

Those of us who were at the first meeting of ITRA at Utrecht will remember that, 30 years ago, play theory was dominated by an increasingly bifurcated debate (Pellegrini et al. 1995). On one side stood educators, followers of Erasmus and Locke, who conceived of a higher intellectual purpose to the playthings of children. Karl Groos's claim that play 'has a clearly defined biological end—namely, the preparation of the animal for its particular life activities' (1901) became a tenet of modern life: gaining strength and dexterity and establishing social bonds between them, these playful acts of the young helped to ensure the survival of both the individual and the species. And if play is the work of the child, then toys and games are the tools through which children acquire the attitudes, knowledge and skills that will be useful in later life. Analyzing educators' commentary on toys in the classroom, Brian Sutton-Smith had argued that the idealization of 'playfulness' had become fundamental to education theories of development, learning and socialization in the twentieth century (Sutton-Smith 1984).

It was against this pragmatic rhetoric of play that Huizinga (1955) wrote *Homo Ludens* as a culturalist refutation of the modernist idea that human play was constrained by our struggle to survive as a species. In the introduction, Huizinga explained his intent to break with the biological, anthropological and psychological theories of play as energic release, as socialization, as learning, and as social bonding. For Huizinga, all these hypotheses have one thing in common: they all start from the assumption that play must serve something which is not play, that it must have some kind of biological purpose. This pragmatic logic thus devalued play as a cultural force which he insisted always manifested itself as 'free activity' of meaning making—imagining, narrating and re-enacting the human experience through myth, ritual, drama, story, and song—that was performed for its own sake. We should therefore tackle 'the problem of play as *a function of culture proper* and not as it appears in the life of the animal or the child' to counter those materialist theories of play. 'The cultural approach begins,' claimed Huizinga, 'where biology and psychology leave off.' Although Huizinga's focus was on culture, his method was philological. Spanning historical periods and cultures, his analysis of the meanings embedded in the word 'play' theorized play as the dynamic spirit of creativity that underwrote all meaning making. 'Summing up the formal characteristics of play we might call it a free activity standing quite consciously outside "ordinary" life. [...] It is an activity connected with no material interest, and no profit can be gained by it.' The end of play, he concluded, was human liberation from necessity.

Believing similarly that games and toys express something intrinsic to the lived experience of human societies, Roger Caillois (1961: 35) agreed that 'the destiny of cultures can be read in their games—or more precisely in the language they use to discuss their experiences of play.' Caillois side-stepped Huizinga's arguments about play's universality as overly abstract—an artefact of Huizinga's lexical form of analysis. It is not just the language of philosophers that can be used to understand play. The language used by ordinary players was rooted in their experience. Caillois believed that play can be theorized as free, but not formless. Game play, he noted, is 'free only within the limits set by the rules' and the rules are defined and articulated by circumstances of that culture. Rules and conventions embedded in the social practices therefore define and constrain the freedom of play's meaning-making process. The language of play thus provided the games researcher with a cultural field that could be read as meaning making produced in very specific historical contexts about the experiences underwriting different expressions of playfulness.

Always a celebrant of the Platonic axiom that 'life must be lived as play,' Brian Sutton-Smith also believed that cultural analysis of play discourse provided an antidote to the bifurcated theories of play as either instrumental or transcendent. Sutton-Smith's research was guided by watching and talking to children engaged in 'folk play' which had highlighted a transgressive element lodged in the stock of cultural knowledge that gets mobilized every time a group of children sit down to play Tag or King of the Castle. It was not just that play was accompanied by language production but 'what was amazing here was how much shared knowledge there was across this group of children of the play forms of all the other children.' Rather than lexical analysis, he insisted that play culture needed to be studied as an organic practice of situated 'meaning making' which, in humans, was characterized by a multidimensional form of social interaction. By seeing play as social communication practiced by players in the context of a specific material culture, Sutton-Smith set out to refocus psychological research into modern play cultures on this paradoxical communication practice rendering the debate between levels of cultural and biological analysis moot.

Although Sutton-Smith can rightly be seen as the godfather of ITRA, if toy researchers are to have a patron saint I would nominate Pieter Brueghel. In his painting *Children's Games*, the whole medieval community seems to give itself over to play with Huizingian abandon. In this vision of a community at play, games seem to be a free and voluntary activity isolated and protected from the rest of life. Work has stopped and ordinary life has been disrupted. Agreeing with Huizinga, therefore, one can conclude that games seem to be 'an occasion of pure waste: waste of time, energy, ingenuity, skill, and often of money.' But Breughel's painting also exemplifies the ethnographic researcher's approach to the material culture of play. Breughel's vision bears witness to play as both the spiritual force of culture making and also as an annotated inventory of specific 'games' sometimes based on material objects (balancing, balls, sticks, etc.) arising in the specific social and historical context of medieval Flanders. In Brueghel's painting, we see a medieval play culture as an embodied social interaction in the village square as a collection of 80 situated specific activities that are both expressive of the play spirit and materially constrained in their social practice. The material culture of play afforded the cultural researcher with a vista of play possibilities written as embodied discourses that get mobilized by the specific community of players in the context of particular social circumstances.

In *Toys and Culture* (1986) Sutton-Smith similarly argued that toy-play activities were not just acts of imaginative social interaction with objects, but paradoxical communicative practices embedded in a broader cultural context that articulated and underwrote the possibilities of playful self-expression. His own cultural analysis focused on the conflicting rhetorics and ideologies that characterized play discourses in the contemporary period—including those of players, play theorists, educators, and marketers. Beyond the ethnography of playfulness, toys signified the values and cultural traditions surrounding child-rearing, education, and human psychology that encourages and legitimizes specific play practices. As Gilles Brougère (2003) later put it, to the playthings researcher 'the toy is more than an object. It is a system of significations and practices, produced by those who distribute it and those who use it, either when giving it or when playing with it.' And here we see the advantage of Brian Sutton-Smith's idea of focusing on the material culture of toy play. Regardless of the specific circumstance of their creation and use, modern playthings are socially constructed artefacts produced in particular historical circumstances of rapid industrialization in a market-driven society which produced thousands of new playthings every year.

The Play Medium Is the Mass-Age

Of course, Sutton-Smith was not the first to notice the industrialization of modern play. Noting the expanding appeal of sports, games, and toys in the modernizing world, Marshall McLuhan (1964) proclaimed: 'If, finally we ask, "Are games mass media?" the answer has to be "Yes". Games are situations contrived to permit simultaneous participation of many people in some significant pattern of their own corporate lives' (p. 210.) As McLuhan went on to explain, games embody a double meaning making, for 'as media that communicate specific cultural values and sentiments' they consolidate social experience. Games therefore serve to valorize the act of play generally. Yet, as 'media of interpersonal communication,' toys and games were also associated with the 'self-expression of players.' Toy and game play media, stood at the juncture of the social and psychological domains of cultural experience.

Seen as paradoxical communication media, both the encoding and decoding moments of toy play are rendered researchable, though not by similar methodologies. As models of our world, all toys are consciously designed as symbols—they point in some intentional way to the known social world, to specific worldly events, situations, objects, or processes. Toy design, and the promotional discourses that sold them on TV, therefore, could be read as facets of mediated communication systems which could be analyzed by isolating the ideological complexities inscribed by the promotion of playthings. In our rapidly modernizing toy world, advertising particularly provided a symbolic window into the changing human roles, rules, relations, and values that are invoked by children in and through playing with them. But at the same time, toys and games are also 'things which can be played at or with.' They set play in motion within a bounded imaginary space which gets negotiated and transgressed by players as they engage with toys and each other. The language of players also provides evidence of this meaning-making paradox.

Inspired by McLuhan's media analytic approach, I set out to explore this expanding zone of our mass-mediated culture as a bridge between two cultural fields of playful 'meaning making' in a market society—the 'encoded' social communication designed into them by the toy makers and marketers and the 'decoded' personal meaning actively constructed by players as the toy's meaning is transformed in imaginative play enactments. As a cultural historian and play media researcher, I wanted to better understand this bridge in playful social communication. The historical account that I provide below portrays the changes in play marketing in three historical phases loosely underscored by changes in market dynamics defined by the industrialization of playthings, the marketing of play media, and the digitalization of play media.

The Industrialization of Modern Play Media

In a little noted passage of *The Wealth of Nations,* Adam Smith comments on the link between playfulness and invention. Smith notes that 'in the first fire engines, a boy was constantly employed to open and shut alternately the communication between the boiler and the cylinder, according as the piston either ascended or descended. One of those boys, who loved to play with his companions, observed that, by tying a string from the handle of the valve which opened this communication, to another part of the machine, the valve would open and shut without his assistance, and leave him at liberty to divert himself with his play-fellows. One of the greatest improvements that has been made upon this machine, since it was first invented, was in this manner the discovery of a boy who wanted to save his own labour.' Playfulness, in Smith's account, was not only the force underwriting the lad's playful self-expression on the job but, through its material implications, an impetus to industrial innovation with profound consequences for socioeconomic change.

As a medium of social communication, innovation in toy design and distribution has also significantly transformed the play cultures of children. Gradually, the European craft toy making of the late-nineteenth century was supplanted by an equally impressive twentieth-century surge in industrialized toy making associated with the rise of the US toy industry. Although European train sets (of wood and metal) embodied the progressive elements of industrialization, too, it was the 22-year-old Joshua Lionel Cowen who created and successfully marketed a battery-powered train engine, the Lionel Train, which became the heart's desire of many young lads. John Lloyd Wright (son of the famous architect) invented Lincoln Logs, a construction set built on the model of a traditional log cabin—which, like the Cowboy and Indian toys, speak of the colonial origin myth of the American nation. Charles Pajeau, a stone mason, developed another construction medium, Tinkertoy, after observing children playing with pencils and spools of thread. Perhaps one of the most telling examples is Charles Darrow's popular board game Monopoly, which was mass-marketed in 1936 by Parker Brothers at the height of the recession. Indeed, as Gary Cross (1997) has suggested, after World War I not only did American toy industries take over the manufacture of traditional toys—pull toys, dolls, plush, and models—but also 'invented' and then marketed hundreds of new toys rooted in the American experience and ideology. One might argue that, especially in the USA, craft toy making was being transformed into an industry which designed, made, and sold toys in department stores just like other consumer goods. With Eric Clark (Clark 2007), we may simply appreciate the restless creativity of the personalities that propelled the rapid expansion of the American toy industries in the twentieth century.

Cultural historians have long argued about the emergence of 'modern childhood' in the mid-nineteenth century. Although toy making was a long-established aspect of the premodern economy, the instrumental view of 'play as learning' helped justify the acceptance of toys as integral to a child's healthy development. Froebel's 'gifts' were an early example of how play could be incorporated within the curriculum of the kindergarten. The idea of play's benefits was transformative: by the twentieth century, not only did playgrounds and sports fields dot the urban landscape, but toy makers had begun to innovate in the design and production of educational playthings to supplement the traditional toys. Edwin Binney's invention of the crayon allowed children to draw and color inexpensively, while Playskool developed a series of puzzle-like learning toys specifically for the nursery. Sand play was introduced into the nursery and puzzles supplemented books. Especially after WWII, educational theory embraced the idea of social, emotional, and cognitive benefits of skill, construction, and learning games—incorporating a myriad of playthings into the very heart of early childhood education. Peter K. Smith (1988), a founding member of ITRA, documented how the 'play ethos' fostered a toy-based ecology of the nursery school.

But this enthusiasm for the pantheon of modern toys had its critics. Roland Barthes (1972) claimed that mass-merchandising had undermined the magical relationship struck between the child and the toy. The problem with mass toy production, Barthes argued, can be recognized not only in their forms, which are all functional, but also in their substances. When contrasted with the hand-crafted artefacts of wood and metal, the plastic toys marketed today seemed to him graceless, lacking the charm and artisanal qualities of folk toys from previous generations. He laments the disappearance of wood, for example, which he believes is an organic material which does 'not sever the child from close contact with the tree.' He also condemns the plastic and metal toys of today for destroying the sweetness of human touch and sensuality of traditional playthings. He prefers the toys of the pre-industrial marketplace made from 'familiar and poetic substances' and in conditions of less alienated labor. In the diverse entertainments provided by modern toys—the kitchen utensils and baby dolls, the trucks, electric trains and car washes, even the Lego bricks—Barthes finds a system of commodities that is 'meant to produce children who are users, not creators.'

The Marketing of Play Media

As Clark noted, the US toy makers not only innovated in materials, production methods, and play values—but most of all in advertising and marketing discourses, which also became deeply etched into the design, packaging, and selling strategies of modern toys. And the pace of innovation in the US toy market accelerated as toy merchandisers realized that play media, like other commodities, could be sold to parents by advertising. As Gary Cross has shown, magazine ads directed to parents articulated the importance of playthings not only for kids' learning but also for their character development and psychological adjustment to modern times (Cross 1997). By the 1950s, play media had a dual life within the industrial market—as playthings which promoted playfulness and as playful commodities that were laying the foundation of the leisure industries.

After WWII especially, the invention of television afforded the toy industry a chance to further innovate in direct-to-child toy marketing. Building on arguments about advertising innovation in America generally, I set out in *Out of the Garden* (Kline 1993) to provide a detailed historical account of the evolving mediated market system in which the production, distribution, and consumption of playthings took place. Advertising toys directly to children on TV marks a turning point in the toy industry and the beginning of a belief in the effectiveness of marketing. *Davy Crockett*, a popular Disney-produced television show, demonstrated the potential impact of TV marketing directly to children. Soon kids were sporting

Bowie knives and coonskin caps across America. Mattel's innovative marketing of the Barbie doll on Disney's *Mickey Mouse Club* program with an ad that portrayed Barbie as a real teen model provided new impetus to child-oriented advertising campaigns. Commercialization of children's TV meant that toy advertisers provided almost half the financial base for the production of TV cartoons for children. Propelled by advertising interest, the TV industry itself innovated in the postwar years in children's programming, funded largely by toy and food advertising directed to children.

But TV was not the only medium for promoting toy sales. The first *Star Wars* movie was not only surprisingly profitable, but the spinoff merchandising, including licensed manufacture of the popular characters and technologies that populated this imaginary universe, were a wake-up call. The deregulation of children's TV marketing during the early 1980s in the USA allowed the children's media industries to further explore the symbiotic relationship between toys and TV narratives. They found that, beyond their ads, programs which visualized the back-stories of their action heroes—Darth Vader, He-man, Transformers, Ninja Turtles—could act as the flagship for a flotilla of branded merchandise from toys to lunch boxes. Synergistic action-toy marketing became the driving force behind a boom in children's goods generally, but a ten-fold increase in toy sales.

Marketing and Its Discontents

I began my career as an analyst of children's culture in 1984 partly motivated by my son's fascination with those 'action toys' advertised on children's TV. I was both intrigued and alarmed by my son's deep fascination with his vast host of action toys—the plastic superheroes and robots that were heavily promoted on TV. Concerned because the play narratives he orchestrated were ritualistic recreations bounded by the characterological framework of the TV characters. But I was also impressed by the range of popular cultural knowledge he derived from TV, and the imaginativeness with which he scripted the play battles, rescues and social moralities of this derivative imaginary world. Regardless, the intensified link between my son's playful self-expression and mass-mediated culture were on perpetual display in my living room. As a parent, I was not alone. In *Out of the Garden*, I therefore also noted how the changes taking place in child play also provoked a broadening critical reflection on the changing social conditions in which playthings have become mass-produced and distributed in the mediated marketplace (see CCFC).

Barthes was not alone in his anxieties about the cultural values being projected by marketers into children's culture. The American toy and game industry has been subjected to wide-ranging social criticism, including feminist diatribes against Barbie's impossible body and housewifely roles to educators' concerns about children's war-play rituals and the urbane brutality of *Grand Theft Auto*. So, too, the moral panics about the banality, sexism, and violence permeating children's media drew attention to the importance of toys and games in children's lives. No longer the cute baby dolls and toy soldiers of the Nutcracker ballet, the imaginary worlds conjured by toy and video game marketers celebrated the increasingly consumerist lifestyles and geo-political realities of our troubled global market society.

Given their popularity with young children, the militaristic superhero TV series of the 1980s have repeatedly been singled out. Teachers warned that many boys were so fascinated with these new televised superheroes that they were assimilating the aggressive back-stories and re-enacting them in a highly ritualized form of 'war play' (Carlsson-Paige and Levin 1987). Unlike the toy truck or baby doll, promotional toys retained no direct correspondence with familiar objects in the real world, but rather to dramatized crystallizations of fictional fields—scripted to refer to mythic worlds-in-conflict. Yet, as Umberto Eco has noted, toy soldiers are a very old play medium which refer paradoxically both to the real world in which soldiers are associated with violence and killing, and to the imaginary world of playful possibility. In play, the soldier takes on a new transcendent meaning by becoming the embodiment of the players' ability to express their inner needs, conflicts, and ideas with toys. The excitement that sometimes accompanies young boys' war play has led many daycare facilities and kindergartens to banish these toys from schools—yet the same exuberance has also been heralded as a sign both of the vitality of meaningful 'free play' and of the value of the play ethos (Goldstein 1998).

Ironically, these evolving critiques of commercialized media are so well known that they have provided the back-story for a recent revival of traditional toys. In the trilogy of films called *Toy Story*, the Disney Corporation has articulated an entertaining critical reflection on the special bond between children and toys, forged before the marketers got involved. In the first film, *Toy Story I*, the transformation of TV character marketing of the 1980s is the backdrop for the disruption of the play universe of contemporary children. Andy is a typical suburban kid with a special affection for his favorite toy, a 1950s-style cowboy rag doll, named Woody.

Woody belongs to the universe of traditional play values grounded in the assumption that the meaning of a toy arises from its owner's devotion to imaginative play—rather than the themes inscribed into its commodity form by the mediated marketplace.

Woody is the leader of a similarly tradition-minded collection of toys dating from the 1950s, including a Mr and Mrs Potato Head, Little Bo Peep, a Slinky dog, and a host of small lead toy soldiers. The drama unfolds when a brand-new Buzz Lightyear toy is introduced into the playroom, which both rivals Woody in Andy's affections and vies with his homespun style for leadership of the gang of toys. Metaphorically speaking, Buzz threatens to replace the backward-looking, but good-natured frontier myth with a futuristic high-tech bravado encapsulated in his hard-wired slogan of 'to infinity and beyond.' The deeper problem for traditional toys, however, is that Buzz takes himself seriously: he doesn't understand that he is *just a toy*. His pre-programmed scripting lacks both imagination and an understanding of the mission of all toys—to be played with enthusiastically. Buzz thus represents the revolution in play values wrought in children's media culture. Yet gradually the friendship develops between Woody and Buzz into a lasting cooperation through their mutual interest in stimulating Andy's playful imagination. The seemingly opposing play values can be resolved if Buzz learns the code of the toy world.

But in *Toy Story II*, Woody's identity as a traditional toy is itself called into question. While trying to rescue Wheezy the Penguin (whose voice box failed) from the lawn sale (the fate of toys no longer loved by their owner), Woody is stolen by an unscrupulous owner of Al's Toy Barn who has now completed a 'collectors' set' of promotional toys to be sent to a toy museum in Japan. Taken to the inner sanctum of the toy store, Woody has an identity crisis. He is only a damaged rag doll supplanted in the affections of Andy by the high-tech playthings of today. Once Woody's arm is repaired, he is ready to be packed for the museum where he makes a fabulous discovery: he is not just a toy like others after all, but a renowned and highly valued antique toy—the prototype TV promotional toy. His character was the centerpiece of one of the 1950s' most popular TV shows with a theme song, horse (Bullseye), and two side-kicks—Stinky Pete the aging prospector, and Jessie the yodelling cowgirl. But that is not all, for as one of America's original TV toys, he has left his impression on a universe of consumer paraphernalia from children's cowboy hats to themed record players. When Buzz and gang arrive to rescue him, Woody has to make a difficult choice: to return to the true meaning of playtime through enriching the imagination of children, or to be immortalized as a collector's commodity.

Although Woody chooses to remain a play companion, the inevitable end of all toys happens in *Toy Story III*, as Andy, now heading off to college, must rid himself of the accumulated playthings of his childhood. As children grow up, toys lose their special role of sparking the child's imagination. For as adults, we must resign ourselves to a regime of work that leaves little scope for truly imaginative play. In the metaphor of the overcrowded toy box, we have a powerful allegory for the crisis of consumer culture that has embraced abundance over creative imagination. Without children to play with, toys are reduced to simple commodities—decorations, collectables, memorabilia—or waste. Their special meaning granted by the child's love of play evaporates and they become clutter, trophies, nostalgic commemorations of the simple pleasures that one knows in childhood. But they are no longer playthings. The only way to be true to their mission is to find another child who will play with them. But here the story gets darker, for the gang of toys are recycled to a nursery where children have lost the ability to play imaginatively. In this ironic twist, the abundance of toys in the market is the undoing of a sustainable play culture. The children are merciless with their toys: they smash, break, throw, kick, and crush them, rather than play imaginatively. A better choice for Woody might have been the toy museum, after all. Like all Disney films, it has a happy ending. Woody and his gang find the one imaginative child at the nursery. Andy goes to university, assured that his toys have not been renounced, but have found an appropriate place to keep that potential 'gleam of freedom' alive. Yet the dystopian vision of the nursery in the digital age lingers in my mind like the smell of burning leaves on an autumn evening.

Digital Play and Synergistic Media Culture

In *Out of the Garden*, I tried to show how deregulation of children's television helped to expand the market for promotional toys by forging marketing synergies between visualized story-telling and play media. But in that work, I ignored an important phenomenon impacting children's play media in the 1980s—namely, the technological synergy linking the TV set to a game console. In *Digital Play* (Kline et al. 2003), I set out to make up for this oversight by analyzing the many ways that innovations in communication technologies continued to underscore the expansion of play media. What I concluded, however, was that digital gaming flourished from innovations in the business models and promotional strategies as much as from the technological inventions underwriting the expansion of video game markets from 5 billion in 1985 (when the industry separated from the toy and game industry) to the 80 billion-dollar industry it is now. The trajectories inscribing the evolution of play media were technology markets and culture. Dubbed the three circuits model, a historical framework was developed as a temporal map to guide an exploration of the interplay between technological innovation (i.e. chips, interface design, 3D graphics and sound, joysticks, smartphones) and changing play practices (game design, new genres, online gaming, gaming olympics) in the constantly dynamic digitalizing marketplace. The new media of pods, pads, and smartphones have since provided a second impetus to this expansion of a digitally synergistic media system, resulting in new gaming forms and practices.

My work has emphasized the role that marketing innovations play in the transformations of contemporary children's culture. I can find no better example for my argument about why toy-play researchers need to be mindful of marketing synergies as much as technological innovations than the case of Lego's recent return to profitability. As Stig Hjarvard (2013) has argued, the recent mediatization of Lego demonstrates that even this quintessentially modernist construction toy can be promoted as an 'imaginary invocation' of mediated fantasies. Originally conceived as a simple brick with which children could 'endlessly' reconstruct miniatures of their changing world, Lego now includes computer-controlled versions of robotic buildings, sexually coded construction sets, and a *Star Wars* video game which references its redesigned toy characters, as well as its own film celebrating Lego values to rival *Toy Story*. In our media-saturated society, toys are no longer 'just playthings.' But neither are they 'just commodities.' Lego is no longer a simple construction toy, but a road map to digital play culture—and the discourses that embedded their use. I have since wasted a lot of ink noting how the same instrumental rhetoric that justified toys as learning technologies came to legitimize the use of computers, robots, and digital devices of all kinds in today's classrooms and living rooms (Kline 2003a, b). The promotional rhetoric surrounding the sale of digitally enhanced toys is familiar. Speaking of the burgeoning sales of tech toys that took the 2016 Toy Fair by storm, the industry declared: 'Parents and educators appreciate these toys because they help prep kids for school by building important spatial, reasoning, critical thinking, and problem-solving skills' (TIA 2016). In this respect, I assert that research into the material culture of play needs to be reminded of the market's continuity, as much as of the technological innovations that are shaping the play cultures of the future.

In a chapter entitled 'Not Just Playthings,' a recent market research report from the NPD Group notes that innovation in American toy design and marketing has helped this industry span the globe (Gifts and Toys 2009). Over 2000 new toys enter the marketplace each year, and the average child in the affluent west can own over 300 toys. The 'not just' reference in the report chapter's title refers to the now accepted idea that the toy and games sector is not to be taken lightly—either culturally or economically. While North America still represents 30% of a worldwide toy market, it is closely followed by Europe at 29%, and Asia at 27%. This report estimates that the global toy trade is worth upwards of US $80 billion. This figure approaches $160 billion if video games are included in the estimate, and a lot more if the whole communication sector is taken into account. Household spending on communication has more than doubled over the last 20 years in developed countries, as digital communication fuels growth of the 'entertainment economy.' Communication media—including toys, video games, and smartphones—are no longer luxuries, but the primary interface between human beings and the engine of economic expansion. For this reason, I am not the only one who believes that digital playthings are the cultural foundation of an expansionary entertainment economy—not only because it is profitable and growing steadily, but because play media for children is one expense that consumers refuse to cut back on, even in a recession.

McLuhan once explained that, in the electronic era, media analysis is best served by 'studying media as cultural environments.' I have interpreted this to mean that if the 'medium is the mass-age' then the message of digital culture is synergistic marketing communication. The hyper-commercialization of play culture has not only magnified toy and game industry profits, but also spawned new critiques of commercialization that augmented the standard critiques of war play and sexist role modeling that agitated the critics of the 1980s. Again, confirming prior insights into mediated market environments, my work has highlighted how, once again, synergies between technological innovation and changing cultural values in online communities have underwritten growth of digital play culture rather than transformed it. As McLuhan foresaw, the prior play media have defined the content of new ones. According to a recent NPD report, American toy sales are arising from the ashes of recession with 6% growth in sales: 'Out of the 11 super categories within toys, eight of them posted gains, with Action Figures and Dolls experiencing the highest dollar growth' (*Gifts and Toys*, December 2016). So too, in the new millennium, public opprobrium for the dystopian commercial spectacle has shifted online to the websites and social media platforms that provide the promotional front end for children's trans-mediated synergistic marketing of today (CCFC 2015). For the next generation of playthings researchers and children's culture critics, advergaming and social media have provided an ample challenge for the future of play media analysts. Pace Faulkner.

Ironically, some early digital play enthusiasts, Douglas Rushkoff and Sherry Turkle for example, are now engaged in critical commentary on the future of play culture. At least their commentary on what is now called 'social media' is more realistic about the relationship between technological innovation and cultural change in mediated markets: the same social media that enable gamers from around the world to work as teams also allow terrorists to recruit and spread hatred, children to bully and insult, and cyber criminals to steal identities (Kline 2015). Play researchers should no longer think idealistically about toys as isolated media, but must, as Sutton-Smith insisted, think critically about the intersecting, and often competing, material discourses that inscribe a toy's paradoxical communication. As play media researchers, we should be aware that, in the contemporary mediated marketplace, toy design and marketing have been integrated into the synergistic media environment which is laying the foundation of both the future entertainment economy and children's culture.

References

Barthes, R. (1972). *Mythologies*. New York: Hill and Wang.

Brougère, G. (2003). *Jouet et compagnie*. Paris: Archambault.

Caillois, R. (1961). *Man, Play and Games*. New York: Free Press.

Carlsson-Paige, Nancy, & Levin, Diane. (1987). *The war play dilemma: Balancing needs and values in the early childhood classroom*. New York: Teachers College Press Columbia.

CCFC. (http://www.commercialfreechildhood.org/).

Clark, E. (2007). *The real toy story: Inside the ruthless battle for America's youngest consumers*. New York: Free Press.

Cross, G. (1997). *Kids' stuff: Toys and the changing world of American childhood*. Cambridge MA: Harvard University Press.

Gifts and Toys. (2009). *Toy sales gain ground in global markets*, Dec, July 29.

Gifts and Toys. (2016). NPD: Toy Sales Up 6% in Q1 May 4, 2016. http://www.giftsanddec.com/article/531196-npd-toy-sales-6-q1.

Goldstein, J. (1998). *Why we watch: The attractions of violent entertainment.* New York: Oxford University Press.

Groos, K. (1901). The theory of play. In *The play of man* (Elizabeth L. Baldwin, transl.). New York: Appleton. Chapter 8, pp. 361–406.

Hjarvard, Stig. (2013). *The mediatization of culture and society.* London: Routledge.

Huizinga, J. (1955). *Homo Ludens: A study of the play element in culture.* Boston: Beacon.

Kline, S. (1993). *Out of the garden.* London: Verso.

Kline, S. (2003a). Learners, spectators or gamers in the media saturated household? An investigation of the impact of the internet on children's audiences. In J. Goldstein, D. Buckingham, & G. Brougère (Eds.), *Toys, games and media.* London: Routledge.

Kline, S. (2003b). Toys as communication. In A. Nelson, L.-E. Berg & K. Svensson. *Toy research in the late twentieth century. Part 2.* Halmstad University, Sweden. ISBN 91-974811-2-2.

Kline, S. (2015). The ends of history and the tyranny of the algorithm. *Kinephanos:* http://www.kinephanos.ca/2014/algorithm/.

Kline, S., Dyer-Witherford, N., & de Peuter, G. (2003). *Digital play. The interaction of technology markets and culture.* Montreal: McGill U. Press.

McLuhan, M. (1964). *Understanding media: The extensions of man.* New York: McGraw-Hill. http://www.eetimes.com/document.asp?doc_id=1329041.

NPD action toys gain ground. http://www.giftsanddec.com/article/531196-npd-toy-sales-6-q1.

Pellegrini, A. (1995). *The future of play theory: A multidisciplinary inquiry into the contributions of Brian Sutton-Smith.* Albany NY: SUNY Press.

Smith, Peter K. (1988). Children's play and its role in early development: A re-evaluation of the 'Play Ethos'. In A. D. Pellegrini (Ed.), *Psychological bases for early education* (pp. 207–226). New York: Wiley.

Sutton-Smith, B. (1984). Text and context in imaginative play. In Frank Kessel & Artin Goncu (Eds.), *Analysing children's play dialogues: New directions for child development, no 25.* San Francisco: Jossey-Bass.

Sutton-Smith, B. (1986). *Toys and culture.* Boston: Gardiner Press.

TIA (Toy Industry Association). (2016). http://www.toyassociation.org/PressRoom2/News/2016_News/Top_Toy_Trends_of_2016_Announced_by_Toy_Industry_Association__TIA___the_Official_Voice_of_Toy_Fair.aspx#.V4fX0jkrJcw.

Author Biography

Stephen Kline is a Professor Emeritus in the School of Communication and Director of the Media Analysis Laboratory at Simon Fraser University, Canada. He is the author of *Out of the Garden* (Verso 1993) and *Digital Play* (MQUP 2003) and *Globesity, Food Marketing and Family Lifestyles* (Palgrave 2011).

CHAPTER 3

Toys: Between Rhetoric of Education and Rhetoric of Fun

Gilles Brougère

Are toys educational? What relationship can there be between an object and the act of learning? These questions often surround the discourse regarding toys and are caught in a tension that shows how such questions are poorly formulated, even meaningless. Indeed, the contradictory answers to these questions reflect two ideas (which can sometimes happily meld together): all toys are educational; certain toys are educational (as indicated on the box). It is this 'ordinary' discourse that one finds on Wikipedia[1] under the entry 'educational toy':

> One could make the argument that an educational toy is actually any toy. Most children are constantly interacting with and learning about the world. This definition is ultimately too broad because one could make the same argument about a rock or a stick as it is not uncommon to see a child play with almost anything nearby.

A first version of this text was presented at the International Toy Research Association (ITRA) Congress in Bursa, Turkey, in July 2011. Some internet pages that are mentioned have since disappeared from the web; one of the toys used for the analysis is no longer marketed by Fisher-Price. These changes related to the renewal of ranges and manufacturers' communication strategies do not, however, call into question the analysis because, as we will show, we find other pages on the Internet and other toys that can support the same analysis.

G. Brougère (✉)
Université Paris 13—Sorbonne Paris Cité, Villetaneuse, France
e-mail: brougere@univ-paris13.fr

L. Magalhães and J. Goldstein (eds.), *Toys and Communication*,
https://doi.org/10.1057/978-1-137-59136-4_3

> The difference lies in the child's perception or reality of the toy's value. An educational toy should educate. It should instruct, promote intellectuality, emotional or physical development. An educational toy can teach a child about a particular subject or can help a child develop a particular skill. The key difference is the child's learning and development associated with interacting with the toy.

It is unclear how a toy could educate or teach. But this poorly phrased question did not prevent a sector from developing based on this association between toy and education, thus enabling the development of rich rhetoric.

Two Discourses: Everything is Educational; Only Certain Toys are Educational

The idea that all toys are educational refers to the idea of play activity as educational a mythical vision of play that should be examined as we have done, along with other authors (Brougère 2005; Sutton-Smith 1997). If play allows one to learn, to develop, to blossom, to grow according to the type of discourse, even the trends of the moment, there is no need to distinguish toys and educational games that can be criticized either by their commercial aspect, because of their use of the educational argument to sell, or by their closed and limited facet from a play perspective. Any toy, provided that it allows effective play, is in fact educational, or the question does not arise since it is not the toy that is educational, but the activity itself, whether it takes place with a toy or any other object picked up and used by the child. Non-educational toys are those with which one does not play or plays poorly. This discourse can be readily critical with respect to toys that are too closed, not creative enough, which is a way to reintroduce the distinction between toys that are more educational than others on another level. But one would simply call them 'good toys' rather than educational toys. This vision, critical regarding a marketing of education grafted onto the toy, is consistent and widely shared by educators, but is based on two assumptions that it is legitimate to question: play is educational in its essence; there can be better toys than others, related to their playability. Indeed, one can do things other than play with a toy (collect it, establish an emotional relationship, accumulate a treasure) and the actual quality of a toy refers to the affordances it displays (what it offers in terms of action to a child) and these are in no way universal, but depend on the child, his or her interests, play culture, tastes.

The second option, the existence of educational toys, is rather a historical, social and commercial construction that has seen, especially through the use of objects—e.g. Fröbel's gifts which are not toys (Soëtard 1990; Brougère 1995), games, toys in school or preschool situations, and the formation of specific objects called 'Educational toys/games' which have passed from the school space to the family. Gary Cross (1997) shows how the company Milton Bradley has built its success on the transformation of the Fröbelian material for family use. The cubes are modified in a way that is little consistent with the ideas of their creator (for example, by adding letters) and have been successful, most especially in a context marked by the Protestant worldview that prefers educational investment (or its appearance) to the pure joy of playing. In France, the company Nathan knew how to lean on its reputation as a supplier of educational materials in preschool to offer educational games for families. This question of the educational toy experienced a significant development with the emergence in the United States since the 1920s of new toys for young children. Thus, Playskool was inspired by 'the equipment used by psychologists in their ten key tests and transformed them into toys' (Musée des arts décoratifs (MAD) 1977). Beyond their differences, these two visions of the relationship between toys and education lead us to consider that they fall under rhetoric or rather several forms of rhetoric, discursive or visual.

All Toys Are Educational

In the first case, this is rhetoric in the classical sense, fairly close to rhetoric unique to discourse on play that Brian Sutton-Smith (1997) analyzes. Among these, he mentions the rhetoric of progress that unfolds around the idea of the role of play in the development of the child. It is this rhetoric which gives rise to a specific development, the historical emergence of which is described by Michel Manson (2001). From a frivolous object that does not merit any interest, the toy becomes a valued object, which deserves to be considered and promoted. It is the development of arguments that highlight the learning (then the development) that the toys would facilitate, be it that of social norms (the doll as an educational support for the role of a mother) or, more recently, of fundamental motor or cognitive skills for very young children. Through this rhetoric, toys, or some of them (the good toys), are legitimized by a discourse that shows their full value without this being related to the search for the slightest scientific evidence. These are not hypotheses that should be verified or questions to be elucidated, but a set of arguments repeated author after author as evidence and permission to value objects and areas of experience. They may be related to negative rhetoric, that which emphasizes the negative aspect of other games or toys, the bad ones, without educational value and which vary depending on the authors. Thus, video games can be valued by positive rhetoric by showing their educational value, or be condemned by negative rhetoric such as that which highlights violence or isolation.

This rhetoric unique to this discourse comes from a dynamic of justification and valorization and is not directly related to the product itself. It can leave traces in catalogues or on the boxes, be taken up by companies to promote their objects (to the point of being very far from the reality of the objects themselves). Websites have become essential media for developing such rhetoric around brands or ranges. The Playskool site in the recent past has developed an overabundant discourse on the educational value of play and the brand's toys. Under the explicit title, 'more than play,' Playskool developed a sales pitch promoting play and moreover its products:[2] 'Play is more than just having fun. Play also provides kids with a chance to practice important skills.' There followed an impressive list of skills developed through play where one could find everything from 'anticipation and prediction' to 'control of muscle strength' or 'cause and effect' relationship. To these skills, toys were associated in relation to the child's age. One can give as an example the text that was found regarding the first item:

> As children watch the action of a toy, they begin to recognize its pattern, and over time start to predict what will happen next. Anticipation and prediction is one of the foundations for developing problem-solving skills—once a child can predict future results, they may modify their play to solve problems

The website associated this aspect with the very history of Playskool, thus showing the continuity in the company's rhetoric[3]:

> Playskool was born from a simple idea: what if toys were not just there to entertain young children, and they also incited their minds to learn and grow?

But our commitment to provide quality toys, sources of enrichment of children through play, remains at the heart of the Playskool name.

On the current website of the competitor Fisher-Price, one finds the same dynamic (temporarily) abandoned in Playskool's commercial discourse in favour of a more commercial approach focused on the products and fun. There we find a heading 'age-by-age playtime guide.'[4] One finds the following comments about the various aspects of development: 'Physical: Become a confident walker'; 'Cognitive: Be better at entertaining herself, and more deliberate in exploring'; 'Social and emotional: Show affection with hugs, kisses, smiles and pats.' There is no relationship with a particular toy (except through pictures of children playing with Fisher-Price toys that are not identified); this is a general relationship between a brand and development of the child as Playskool had offered before.

The Educational Rhetoric of the Toy Itself

Beside this rhetoric around the activity and the toy, one can identify another rhetoric, that which the toy itself develops, visual rhetoric (Brougère 2014) which leads the object to contain signs that allow it to persuade, to seduce (Charaudeau 1997) through its educational value. Rhetoric which is addressed, like the previous one, to parents, to adult providers of toys, constructs the association with the child, obvious for a toy, around an educational legitimacy, a rationality that goes beyond entertainment, which is not always sufficient to value the toy, at least for some adults. The younger the child, the more the rhetoric of education is present in and around the toy, for two reasons: first, because the child takes little part in the decision process; and, second, because educational issues, especially before kindergarten or elementary school, depending on the country and the educational system, seem to fall upon the family which must provide a 'preparatory' curriculum without, however, entering into a school dynamic. The toy seems to be the ideal instrument, if one can show that it promotes education while remaining on the side of play, amusement, entertainment. This appears in the very history of Playskool which shows the importance of rhetoric unique to the object itself.

Indeed, it is a matter of using entertainment for educational purposes, which is not very far from education by trickery as Erasmus proposed (Brougère 1995):

> The notion of 'learning through play' was the inspiration of two women who worked for the Schroeder Lumber Company in Milwaukee, United States. Former school teachers, they based their very first Playskool creations on educational tools they had used in class. Their first 'production' was a wooden writing desk with lid filled with supplies for fun learning, such as cubes, pencils and a slate. The first Playskool catalogues contained the desk as well as other sturdy toys such as a folding wooden doll house.[5]

This alternative vision of the educational toy consists of closely associating play objects with the idea of education or development, not around the idea that play is educational, but around signs that allow the educational dimension to be seen in the toy itself. Indeed, the rhetoric of education refers to the desire to make the educational aspects visible, a visibility corresponding to a variety of signs. The simplest and most frequent come from signs strongly associated with school, therefore with learning, in a vision that closely links the two. The use of letters and numbers on a toy makes up the basic rhetoric. Others may associate colors with it as a learning theme, or, for older children, words. One can note the predilection for the world of animals and a distancing from references to the world of daily urban life. The rhetoric of education must be associated with addressing childhood, which favors themes like those of animals and nature. The educational toy appears, above all, as an object that features education, using cultural signs coming from the educational world.

A Toy Telephone for Learning

Let's go a little further in this analysis by taking a concrete example, a telephone offered by Fisher-Price, that has since disappeared from the range. This is the 'Fun 2 Learn Cell Phone' (Fig. 3.1). It belongs to the range 'Fun 2 Learn', very close to the range 'Laugh and learn' which still exists from this manufacturer. Around the object we find rhetoric of justification for the range around the convergence of entertainment and learning:

Fig. 3.1 Fun 2 learn cell phone by Fisher-Price

> ***Children love to learn while they play****—and learn best when they're having fun. That's why* ***Fun 2 Learn™*** *toys bring learning to life, making it more enjoyable with the help of friendly characters, role play activities, music, games, and more.*

It is a matter of blurring the boundaries between fun and learning or, to put it another way, of closely associating rhetoric of education and rhetoric of fun. As for the object itself, it is presented as 'the only cell phone that helps your child learn important phone numbers.'[6] The more specific description, which has since disappeared, on the manufacturer's own website explains how the child can learn the phone numbers by programming them in themselves.[7]

Though the toy is accompanied by a discourse that promotes the educational aspect, one can also track the rhetoric in the object itself. The mobile phone includes three modes, where the child can learn real phone numbers, letters and numbers, and make 'calls' to talk to new friends.[8]

This information and the following come from the instructions for the toy as sold in France. The product has been localized (use of French and local emergency numbers).

The following aspects of the object come more specifically from the rhetoric of education:

- The phone keypad (which implies learning its use presented as learning of emergency numbers, and we can interpret this as learning the use of the things we have to consume or learning the consumption of a cell phone).
- The numbers and letters (with in the latter case a little game): hear the name of the numbers, recognize the letters (in fact, numbers and letters mean school and are thus markers of educational skills, elements of the rhetoric of education).
- The displays on the phone's screen come with the comments: 'Tom the fire-fighter drives a big red truck and can be reached at [the number] 18'; we find on the same level: 'Pizza Paul who cooks pizzas in the oven or cuts them into slices,' or Julie the police officer: 'Police officers help and protect people, I make sure that people respect the law. In case of an emergency call me at [the number] 17, be careful,

goodbye.' (The machine asks the child to reproduce the number but does not provide correction if she or he fails, simply solicits two responses and, if successful, repeats the numbers and says OK.)

Thus, the rhetoric of education is present in the object through numbers and letters on the one hand, the emergency numbers on the other. To this is added a realistic aspect (of course, limited) that refers to a real mobile phone without, however, being a faithful representation.

But what characterizes such a product, faithful in this to the name and essence of the range (Fun 2 Learn), is the presence of a rather strong fun or entertainment aspect.

This is translated by the fact that the phone is an anthropomorphic object, with feet and an image of Gabby appearing on the screen, that it allows one to pretend to call or take pictures, and that it offers games and music.

Though certain signs display what we have called rhetoric of education (shapes, colours, letters and numbers), including being addressed to the child (it is an object intended for the child, it is not a real phone, but an imitation that highlights the supposed learning elements), other signs indicate that it is also a toy, associated with play as amusement and entertainment (anthropomorphism, music, pretense, screen).

Though this phone has since disappeared, the same brand currently offers a learning phone with arguments similar to those mentioned for the previous model. The toy offers 'learning songs and musical ditties,' in addition to a sing-along aid in the form of a smiling puppy, offering an early introduction to role play.[9]

This toy finds its place in the range 'Learning & Educational Toys,' presented as follows on the current Fisher-Price website:

> With the help of Fisher-Price educational toys, children can learn important concepts through engaging and interactive play.[10]

This range includes 119 toys, of which 62 are under the brand Laugh and Learn. These are toys that refer to objects in the real world but have been transformed by way of shape, color, or size.

Fun and Education

This object therefore demonstrates (as it says in the name itself) what we have called the rhetoric of fun (Brougère 2014), a way to address the object to the child in relation to entertainment. It conveys the importance of this aspect, the legitimization through education certainly no longer having the same importance that it once had. One can consider that today the toy is based on recognition of play as legitimate entertainment for the child. It is therefore expected and not only accepted that the toy addresses itself to the child as an entertainment medium; education must, consequently, take on an amusing form.

Such was the dynamic of 'edutainment'. This hybridization between education and computer entertainment or digital game had its (limited) heyday, but seems to have disappointed both users and businesses by stumbling on the difficulty of offering a sufficiently fun experience. Indeed, faced with the development of video games that progressively improved their animations, their play patterns, and their ability to immerse the player, edutainment whose educational lines were sometimes a little too visible had a rather hard time convincing its users. It was easier to convince parents. Analysis easily showed, with a few rare exceptions, that the play aspect was sacrificed to the educational process. All that was needed to realize this was to question the experts—the players, including the youngest ones (Ito 2008).

To address the child, which in the past was closely linked to the idea of a child marked by the act of growing up, learning and developing, is today strongly linked to entertainment, to pleasure, whether it be part of a family relationship experienced not only as an educational place but as a space for mutual blossoming, or within the framework of peer culture that pays little interest to what is given as educational in an explicit way.

As a result, toys (but also other objects intended for children) are part of a rhetoric of fun, whether to seduce the child (and parents) through objects which are recognized as intended for the child (as is the case with toys or candies) or objects that are transformed to be specifically addressed to the child (like certain clothes or novelty stationery).

Even if it includes an educational aspect, the toy must be fun and, by tradition, Fisher-Price pays particular attention to this aspect:

> The founders of Fisher-Price Toys insisted that play should be a happy experience—their toys were cheerful, 'full of humour and ingenuity, sound and action.' But Fisher and Price also believed that their toys should bring more to the child, give him or her the opportunity to feel, touch, arouse his or her curiosity and invite exploration (MAD 1977, p. 16).

The same dynamic can be found in a toy such as Alphie by Playskool with perhaps a stronger educational aspect, around specific, fairly academic knowledge, in the form of questions, and depicted on the object itself (the presence of ABC on the body of the robot). But though the rhetoric of education seems to largely dominate in such a product that has seen different versions and technological changes since its first appearance in the late 1970s,[11] the rhetoric of fun cannot be absent and is present in the appearance of the robot character itself which is the form given to the electronic medium:

> As his buttons light up, he 'sings' and plays music, engaging your child with fun quizzes and games from his activity cards! Oh, and this alien won't make a mess in the house, either—he's got a storage bin for cards and a convenient handle on top of his head.[12]

The educational toy today reflects a hybridization between the two rhetorics mentioned. Less realistic, less oriented toward preparation for adult life, the toy has developed in recent decades in a dynamic of imagination and entertainment (Cross 1997; Brougère 2003). It no longer has the task of preparing for the future or teaching the world to children, but rather, like other entertainment, of offering an alternative space to the real world, where it is possible for the user to develop pleasant experiences for themselves. The educational dynamic had no problem connecting with the preceding vision of the toy; there was convergence around a toy legitimized by its support of becoming an adult. Of course, here it is just justification, children finding in the toy a place of pleasure in which to escape the constraints of the everyday world. No doubt the very dynamic of contemporary toys has gotten closer to the meaning it has for the child, in so far as a hedonistic orientation of society, accepted for adults, has likewise been accepted for children, at the very least in the family space.

However, the desire to educate through the toy, including by trickery, remains present, especially for the youngest children—those who do not, or only partially, choose their toys. The constraint is then that of respecting the two aspects: to remain within the dominant dynamic of the toy and to accept fun, while still opening opportunities for learning—or rather, since it is by no means a matter of assured or verified learning, the rhetoric of education, for the toy, has become acceptable only if it incorporates the rhetoric of fun. The specificity of the toys, and beyond that of games, is to provide fun education, associated with entertainment, or at the very least to display this in the discourse accompanying the toy and in the toy itself, as we have seen. The contemporary educational toy should therefore display signs that show the educational (such as letters or numbers) and the wrappings of fun (such as anthropomorphism, cheerfulness, and music).

One can ask oneself if the change to the Playskool website, less didactic, does not go in the direction of more fun, entertainment. This reflects the search for a balance between the two aspects and oscillations regarding what one should highlight, seeking to conserve simultaneously education and entertainment, which may appear contradictory. How to be fun and educational, how to make serious games, is undoubtedly a central issue of the rhetoric of objects intended for children, at least of those which are not focused on entertainment without any other justification nor linked to an exclusive educational end. It should be educational while still being a toy, which means today to be linked to the rhetoric of fun.

Here it is not a matter of asking oneself whether all this has a learning effect, which comes from another inquiry. Whether the child learns is largely independent of the rhetoric of education. Education is here, above all, a matter of rhetoric, of sign, of image, which is consistent with what a toy is, an image of the world intended for the child (Brougère 2014) and within the framework of a domination of the rhetoric of fun that involves developing objects of compromise, hybrids (with more or less success) that combine education and fun.

The rhetorics that we highlight certainly allow one to address an object to the child (or to the parents for the youngest children), but beyond constructing the recipient child, situating him or her either as a future adult, or as liable to be entertained. The incorporation of the two is undoubtedly one of the characteristics of many contemporary products. It is a matter of seeing that addressing the child is addressing a child marked by age and gender, but also by the values that we accord to childhood. In addressing the child (directly or through the mediation of an adult), one performs the task of defining what a child is. Today it appears as linked to fun and entertainment, including in educational tasks within the family and undoubtedly beyond, at least in certain cultures. The issue of serious games leads us to discover similar dynamics on another level. In this case, it is a matter of implanting in supposedly serious products, often coming from the rhetoric of education, the dynamics which come from video games, therefore largely marked by the rhetoric of fun (Deterding et al. 2011). These hybrid objects incorporating a double discourse can only pose a problem. Does the education go beyond rhetoric, the staging of education through images and signs? Is entertainment always present when the will to educate is present?

Notes

1. https://en.wikipedia.org/wiki/Educational_toy, date accessed October 26, 2015.
2. http://www.hasbro.com/playskool/fr_FR/discover/more-than-play/ date accessed July 12, 2011; content has since disappeared.
3. http://www.hasbro.com/playskool/fr_FR/discover/for-parents/why-playskool.cfm date accessed July 12, 2011; content has since disappeared from both the French and the American website.
4. [http://www.fisher-price.com/en_US/playtime-guide/12-18months/index.html] date accessed October 26, 2015.
5. http://www.hasbro.com/playskool/fr_FR/discover/for-parents/why-playskool.cfm date accessed July 12, 2011; content has disappeared from both the French and the American website.
6. http://www.walmart.com/ip/Fisher-Price-Fun-2-Learn-Cell-Phone date accessed July 12, 2011; content has since disappeared.
7. Ibid.
8. Ibid.
9. http://www.fisher-price.com/en_US/brands/laughandlearn/products/Laugh-and-Learn-Learning-Phone date accessed October 26, 2015.
10. Ibid.
11. Information on the different versions of this toy can be found at the following address: http://www.theoldrobots.com/smallbot18.html, date accessed October 26, 2015.
12. http://www.hasbro.com/en-us/product/playskool-alphie:37C4291A-19B9-F369-10BFEC1BA496D865 date accessed October 26, 2015.

References

Brougère, G. (1995). *Jeu et education.* Paris: L'Harmattan.

Brougère, G. (2003). *Jouets et compagnie.* Paris: Stock.

Brougère, G. (2005). *Jouer/Apprendre.* Paris: Economica.

Brougère, G. (2014). Toys or the rhetoric of children's goods. In D. Machin (Ed.), *Visual Communication.* Berlin: DeGruyter Verlag.

Charaudeau, P. (1997). *Le discours d'information médiatique., La construction du miroir social.* Paris: Nathan.

Cross, G. (1997). *Kids' Stuff—Toys and the changing world of American childhood.* Cambridge, MA: Harvard University Press.

Deterding, S., Dixon, D., Khaled, R., & Nacke, L. (2011). *From game design element to Gamefulness: defining "gamification". MindTrek* 11. Proceedings of the 15th International Academic MindTrek Conference: Envisioning Future Media Environments. Tampere, Finland, 28–30 September 2011.

Ito, M. (2008). Education vs. entertainment: A cultural history of children's software. In K. Salen (Ed.). *The ecology of games—connecting youth, games, and learning.* Cambridge, MA: MIT Press.

MAD [Musée des arts décoratifs]. (1977). *Jouets américains. 1925–1975.* Paris: MAD.

Manson, M. (2001). *Jouets de Toujours.* Paris: Fayard.

Soëtard, M. (1990). *Friedrich Fröbel. Pédagogie et Vie.* Paris: Armand Colin.

Sutton-Smith, B. (1997). *The Ambiguity of Play.* Cambridge, MA: Harvard University Press.

Author Biography

Gilles Brougère is Professor in Sciences of Education at Université Paris 13—Sorbonne Paris Cité, France. He is a member of EXPERICE, a research centre about education and culture.

CHAPTER 4

A Toy Semiotics, Revisited

David Myers

Introduction

The rapid rise of the video gaming industry has paralleled the growing pervasiveness of digital media more generally. Worldwide video game sales and revenues now surpass those of movies (though these numbers commonly include the digital accessories and hardware—joysticks and consoles—that are required to play). Of course, not all digitally transformed games have been equally successful. The initial popularity of arcade games, for instance, has waned in favor of games played on personal computers and in-home dedicated game consoles (Williams 2006)—indicating that the appeal of digital games is a consequence more of their experiential than material configuration.

Simultaneously, over the past several decades, sales of traditional toys and games have lagged in comparison to video game sales (ECSIP Consortium 2013). This trend is most obvious in highly developed regions of the world, but does not seem otherwise restricted to a particular culture, type of game, or type of play.

While traditional games have flourished in their transformation from analog to digital, traditional toys have not equally benefited.

D. Myers (✉)
Loyola University, New Orleans, LA, USA
e-mail: dmyers.loyola@gmail.com

L. Magalhães and J. Goldstein (eds.), *Toys and Communication*,
https://doi.org/10.1057/978-1-137-59136-4_4

This is not to say that, over the past quarter-century, digital media have not had a significant impact on toys and toy production. The USA National Toy Hall of Fame, for instance, now includes the Atari 2600 and Nintendo Game Boy, alongside alphabet blocks and jacks as 'classic icons of play' (The Strong 2013). Yet commercially successful digital adaptations of popular toys and toy forms—e.g. recasting the teddy bear as Teddy Ruxpin (1985)—have yet to supplant their non-digital predecessors as overwhelmingly as, for instance, *World of Warcraft* (Blizzard 2004) and similar Massively Multiplayer Online games (MMOs) have supplanted *Advanced Dungeons & Dragons* (TSR 1977) and paper-and-pencil wargames.

Differences Between Toys and Games

Why do traditional games seem to display a greater affinity with digital media than do toys? There are several possible explanations.

The appeal of toys is more fundamentally *material* than that of games. One of the most enduring categories of toys is the 'plush' (or stuffed) toy category, and the ephemeral nature of digital media might eliminate those tactile pleasures associated with toy play. In short: digital toys are difficult to hug.

Toys are also more commonly associated with younger—and less sophisticated (including animal)—play. Indeed, influential developmental play theorists (e.g. Piaget 1962; Montessori 1967; Vygotsky 1933) have established a continuum of play, extending from child to adult, in which the child's earliest play with body, environment, and objects is a precursor to play with more complex ideas and concepts.

From such a theoretical perspective, flat and/or declining toy sales may indicate that the period of developmental play leading to more advanced and conceptual play has grown shorter, resulting in a decrease in market demand for toys. This cognitive acceleration of youthful play might be a consequence of cultural values and institutions prioritizing early childhood education and socialization, particularly in more highly developed countries—or simply a consequence of more and more easily accessed opportunities to interact with digital media and games.

However, these explanations do not account for the shared—and unique—semiotic properties of toys and games, nor do they offer any clear rationale as to why games have been invigorated by digital media per se. Here, I would like to focus on the semiotic properties of toys and games that might equally explain digital media's synergy with games and their relative immiscibility with toys.

The Semiotics of Toys and Games

> The message for the majority of toys is that they signify some property in the real world (dolls for babies; cars for automobiles) and yet at the same time paradoxically signify that they do not signify what these real objects signify (Sutton-Smith 1984: 19).

Games and toys differ from other, non-play objects in the cognitive experience and associated ludic attitude they evoke during play. Their most unique representational qualities are not immediately evident during initial play. It is only during repeated and *recursive* play that toys and games stand apart from other, more conventional signification processes.

Toy Semiotics

There are two tropes commonly associated with the toy as a signifier: the toy as a cultural artefact, and the toy as *real*.

Initially, many toys derive their appeal from reference to existing objects of cultural significance. This is particularly the case in developed nations and in consumer contexts in which the toy must be purchased prior to its use—or, in some cases (e.g. toy collecting), without regard to its use. Thus, Barbie dolls, Easy Bake ovens, G.I Joe action figures, and other, similar replicas of pre-existing objects convey their appeal through a conventional signification process that makes no necessary reference to play.

However, this component of toy signification is a double-edged sword to their appeal. While Barbie dolls represent generic human females as many dolls also do, Barbie dolls are distinguished from their competitors through their representation of human females in a particular social and cultural context (cf. Pennell 1994). When the social and cultural context—including those values associated and linked by this reference—falls out of favor, the appeal of the toy falls as well. Barbie dolls, for instance, have always been fodder for commentary and criticism based on the representational baggage they carry (cf. the analysis in Fleming 1996: 41ff), and, recently, these dolls have declined in popularity (Zimmerman 2009).

Much research and commentary on toys have been based on this trope of toys as cultural artefacts, influencing our understanding of the origin of toys (Kyburz 1994; Levaniouk 2007) and the current widespread promotion and use of toys in formal—and informal—educational contexts (Best 1998; Verenikina et al. 2003).

However, there is another, contrasting trope relevant to the semiotic function of toys: the toy as *real*. And, despite its fanciful origins, this trope is more revealing regarding the unique semiotic properties of the toy. Rather than an embodiment of culture, this trope positions the toy as an embodiment of *self*.

There are a great many stories and folktales that engage this theme of the toy becoming real. These narratives may be couched in a conflict between a toy-like object and the real world (e.g. Frankenstein, Harvey) or as a transcendent journey from the toy world to the real world (e.g. The Velveteen Rabbit, Pinocchio). Regardless, this trope involves a dramatic opposition between reality and player self (or *ego*).

Despite their material bias, toys are decidedly *not real*—or, more pointedly, toys have no reality *of their own*. The reality of the toy is entirely subservient to the reality of its master (and player). This makes toys semiotic objects of a very direct and immediate sort: representations of player *ego*. Toys serve ego gratification quite directly during play, but also might serve more indirectly in broader social and cultural contexts, i.e. as significations of dominance and status ('whoever has the most toys wins'). And, tellingly, a toy cannot continue to serve this semiotic function should it develop, in imagination or in fiction, its *own* ego. For a toy to remain a toy and an object of play, that toy must remain in the liminal space between what is and what is not, a state kept paradoxically 'alive' by the player's ludic attitude.

In this respect, the semiotic function of the toy as an object of play is in conflict with its function as a cultural artefact. During repeated play—if the toy is engaging enough to prompt repeated play—it attains a ludic identity that resists cultural references: it is a paradoxical object bound neither by reference to pre-existing objects and values nor by any fixed reference to its own material reality (e.g. its plushness or its price).

Game Semiotics

Games share the same paradoxical liminal space with toys, but operate, in contrast, according to *rules*. The consequences of this rules-determined aspect of games is, again, not immediately evident during initial play in which the game player is learning the rules (which also requires, in the case of digital games, becoming adept with the hardware interface and controls).

It is only after the game player has become fully aware of, and, to some degree, habitualized to, game rules that the player is able to voluntarily accept their limitations and enjoy game play as a liminal experience. Over the course of this habitualization process, the values and meanings of the game, like those of the toy, are divorced from their real-world references and thereafter determined solely by their (rules-determined) in-game functions.

For instance, for the novice chess player, chess pieces may reference queens and kings and knights. For the grandmaster, however, these in-game referents are denoted solely by the manner in which they move according to the rules of chess. An in-game object labeled 'Queen' might equally be labeled 'Dog' or 'X' insofar as the rules governing its movement are not changed. In this way, the semiotics of toys and games operates similarly in divorcing values and meanings from out-of-game cultural contexts and references.

In contrast to toys, however, games *challenge* the ego of the player. Any ego gratification associated with games is delayed (during initial play) and unavailable entirely without some effort and resulting mastery of the game. This means that the assignation of values and meaning within the conventional game—i.e. its *semiosis*—is more restrictive (i.e. less 'pliable') than that associated with the toy.

While a toy's values and meanings are, to some degree, shaped by its material form, there is still the potential for a doll to be used as a hammer—and for a hammer to be dressed and used as a doll. These two potentials do not preclude the doll—or the hammer—being a 'toy.' In contrast, in order to play the game, in-game objects must be used (i.e. values and meanings must be assigned) as dictated by the game rules. To do otherwise is, in Suits' (2005: 60) terminology, to become something other than a game player: e.g. a trifler, a cheater, or a spoilsport. All such alternatives to the game player assign values and meanings in the game without regard to the proper rules of play, *idiosyncratically*. During toy play, the ego is gratified in just such an idiosyncratic manner. During game play, the ego is comparatively restrained.

Despite this rules-determined difference between toys and games, these two equally share those circumstances that disrupt their play. Toy play is subverted when the toy becomes too real and the game is subverted similarly. Games that become too real—e.g. The Most Dangerous Game, or Russian Roulette—force players out of the ludic attitude that precariously adjudicates between what is and what is not.

Should 'what is' ever take precedent over 'what is not'—during either game or toy play—, then play is ruined. Ruined games include those life-threatening examples above, but also poorly designed games with rules too simple or otherwise unable to sustain an uncertainty of outcome—e.g. Tic-Tac-Toe (or Naughts and Crosses).

All fixed components of digital games that spin along without any opportunity for player intervention or control—e.g. as either narrative or simulation—likewise preclude ludic play. Should any game component be predetermined in outcome or value prior to play, then there is no challenge posed to the player and no doubt as to what is and what is not: the game simply *is*. A game that simply *is*—i.e. determined by its rules rather than by its play—is incapable of evoking the peculiar semiotics of game play. And yet, while neither game nor toy can survive becoming overly *serious*, it is precisely *the pretense of seriousness* that sustains the liminal mode fundamental to their semiotics.

The Transformations of Digital Media

> It is the motivated action with these symbolic vehicles rather than the medium [sic] themselves that constitute the primary focus
>
> (Sutton-Smith 1984: 19).

How do digital media affect the semiotics of play?

In 'A Toy Semiotics,' Sutton-Smith (1984: 19) calls toys a 'first departure' in 'retreating from [an] object of reference' and subsequently argues that computers promote play in 'more symbolic, less physical worlds' (Sutton-Smith 1986: 78). Digital games can be understood to extend this first retreat of the toy, in recursive fashion, by building a syntactical superstructure atop those 'departing' significations associated with play.

Thus, it might seem that digital media display an affinity with games because both share a similar sort of formality: interacting with software code enables a similar cognitive mode, and perhaps a similar ludic attitude, as interacting with game rules. However, the ability of digital media to embody the rules of the game has not always enhanced those aspects of game play that distinguish it from toy play. Digital media transformations of traditional games have also served to *remove* semiotic distinctions between games and toys, making digital games increasingly toy-like in function and, therein, increasingly capable of a toy-like ego gratification.

Video games sales have infringed not only on traditional toy sales, but on traditional game sales as well. Significantly, digital media have revised traditional games and game play in common and consistent ways. Digital games abbreviate and streamline the traditional game rules learning process, provide more customizable game goals than do traditional games, and, as a consequence, have popularized so-called *casual* and *sandbox* games.

One of the more significant developments in the evolution of digital games has been the extinction of the printed game rules manual. Initially, digital games were released, as were their non-digital progenitors, with text-based manuals. Over time, of course, like all other printed materials associated with the game, these manuals have been transformed from paper to pixels. During that transformation, game rules explanations have grown vaguer and briefer, eventually disappearing entirely. The digital descendants of dedicated game rules manuals are now 'help files'—and community-based wikis of various sorts—that are largely superfluous to initial game play.

Digital game rules are now learned during game play through trial and error rather than through study and reflection, if they are 'learned' at all. The digital game player's habitualized recall of previously played and similar games is often enough to avoid any reflection on game rules whatsoever, removing this obstacle to initial game play. In this transformation, digital game rules have come to increasingly function as more natural affordances, restricting and regulating the assignation of values and meanings within the game without any conscious awareness or intervention on the part of the game player.

Many complex digital games, of course, still require close attention to their rules and, often, a steep learning curve needs to be climbed in order to achieve mastery. Complexity, however, is in the eye of the beholder.

Digital games are distinguished from their traditional counterparts in being able to provide a customizable game-playing experience with adjustable difficulty levels. Game players might dial this difficulty down so low that any challenges provided by the game are illusory. This is particularly the case in those narrative-based games in which the game's winning conditions are secondary to the development of story and character. Without any uncertainty of outcome regarding their inevitable dénouements, narrative games cannot function in accordance with the semiotics of traditional toys and games.

Players must either engage the narrative game as text, in a non-paradoxical way (i.e. as a cultural artefact, with values and meanings fixed by out-of-game references) or in a more paradoxical and typically ludic manner in which the narrative is treated as an object of play (i.e. as both narrative and not-narrative at the same time). This latter type of recursive and subversive play has, for instance, come to dominate play of the well-known 'interactive story' game *Façade* (Mateas and Stern 2003), in which the game's narrative structure becomes little more than a rules-based platform for self-centered play and parody (e.g. Cr1TiKaL 2011).

These common and consistent digital media transformations of traditional game forms—blunting the strictures of game rules and reducing the difficulty of game challenges—have resulted in so-called 'casual' digital games, which require less commitment, less time, and less skill from novice players. Casual games might employ complicated software codes and designs, but the rules associated with their game play are implemented in simple and familiar ways that are quickly and viscerally—e.g. visually (Leja 2000)—demonstrable; and the winning conditions of casual games are more often accomplished through increasingly sustained than increasingly difficult play.

In some respects, the transformation of traditional games to digital and casual games has been similar to the transformation of drama and theater to television sitcoms, and the transformation of opera and symphony to radio pop songs: market and economic conditions have similarly shaped entertainment content intended for mass consumption. In this context, it is not unreasonable to characterize the transformation of games by digital media as a 'dumbing down' of traditional game forms. However, digital media allow for multiple configurations of pre-existing forms, with possibilities for incorporating the traditional as well as the more contemporary and consumer-oriented.

The Future of a Toy Semiotics

> Paradoxical signifiers (it is and it is not what it signifies) appear to lend themselves to pliable representation of larger cultural conflicts
> (Sutton-Smith 1984: 21).

'Pliable representations' are not fixed by a particular cultural context, yet there are aspects of a toy and game semiotics that seem threatened by the transformations of digital media.

Traditionally, games offer a more sophisticated level of play than do toys, one in which the player ego is sublimated in order to evoke a unique liminal experience. When playing a game, the game player must acknowledge, through voluntary acceptance of the game rules, that the player ego occupies the same liminal space and carries the same liminal status as all other in-game objects: the player ego both is and *is not*. The innate human ability to construct (and enjoy) a recursive and self-reflective liminal experience of this sort appears under no threat of digital transformation. However, this is only the most basic semiotic function of play—an *in-game* function. Should paradoxical (i.e. play-based) signifiers be as pliable as Sutton-Smith suggests, then the 'larger cultural conflicts' referenced by digital games and toys bear inspection.

The consequences of play with toys and games have, in the past, been different. Semiotically, since games are less egocentric than toys, games have forced an increased level of self-reflection and, potentially, self-understanding. Digital media do not appear to preclude this semiotic function of in-game play; however, the same characteristics of digital media that have allowed games to become more toy-like have also allowed games to become more *reality-like*.

Prior to digital media, the effect of the rules of the game and the exerted control of these rules were enabled only through a conscious and voluntary decision on the part of the player to learn and abide by those rules. The game was therein a *pretense* (i.e. 'just a game')—and most especially (i.e. most ludically and most paradoxically) a *pretense of seriousness*. Currently, digital games embed their rules within software and code in such a way that these rules govern play through implicit compliance rather than explicit choice. Rather than relinquishing control in order to play, players are immersed in a digitally constructed and maintained *context of control*, in which game rules are indistinguishable from more natural affordances. Game play that pretends to be serious is now threatened by 'serious games.'

The Game as Simulation

In his conclusion to *The Grasshopper*, Bernard Suits (2005) speculates that the future of mankind's endeavors, should those endeavors produce Utopia, is to play games. Yet, in this Utopia, 'although all of the apparently productive activities of man were games, they were not believed to be games [...else] they would have felt that their whole lives had been as nothing—a mere stage play or empty dream' (Suits 2005: 160).

Certainly, digital game technology has evolved to be able to represent the real world more precisely in appearance and more closely in effect—toward the perfect simulation of the holodeck. In parallel, one of the major themes of the modern games studies movement has been to question what is—and what is not—distinctive about play.

For Huizinga (1955) and other early play theorists, play is only possible within its own restricted space: a *magic circle*. That is, '[a]ll play moves and has its being within a playground marked off beforehand materially or ideally, deliberately or as a matter of course' (Huizinga 1955: 10). For more contemporary theorists and critics (e.g. Copier 2005; Malaby 2007), however, this circle of play is more readily and appropriately breached than enforced. Rather than protecting the sanctity of play from real-world intrusions, contemporary theorists have suggested that, in order to broaden the appeal and usefulness of play, play needs to become *more real*—blending toy, game, and reality.

Some Examples

The *Tamagotchi* (Bandai 1996) series of 'virtual pets' was a relatively early, digitally realized blend of traditional toys and games. In its original form, the digital microprocessor in the egg-shaped Tamagotchi device combined the appeal of a toy doll with the immersion of a role-playing game.

The initial market demand for Tamagotchis was extremely strong (though this extreme level of demand was short-lived, see Higuchi and Troutt 2004). And the Tamagotchi—along with many other 'virtual pet' variants—has held a continuous presence in the toy marketplace since the 1990s, long after the widespread social phenomenon associated with the release of the original Tamagotchis had died down.

The quick rise and fall of the Tamagotchi devices can be attributed, in part, to the absolute necessity of the interaction between the original Tamagotchi software and the player (most particularly in those earliest versions that had no pause buttons). The constant level of interaction required proved burdensome for players as well as problematic for those parents and school teachers who found the toy distracting and intrusive on other activities (Bloch and Lemish 1999).

The ego gratification associated with Tamagotchi play was made more appealing by embedding game rules that digitally recorded and objectively realized and valued that play. At the same time, however, player ego gratification was controlled, delayed, and made more difficult by the imposition of inviolable (non-customizable and, perhaps, overly difficult) game rules. In retrospect, the original Tamagotchi designs simulated the mechanics of pet care—including pets 'dying' without round-the-clock attention—*too* realistically. Subsequent designs of digital-based toy-game hybrids have attempted to retain an illusion of reality while selectively omitting those components of reality that make play something other than *fun*: i.e. reality has proved less preferable than verisimilitude.

The original version of *SimCity* (Maxis 1989), as submitted by Will Wright to Broderbund as a game, was rejected: the design did not have traditional game goals. *SimCity*—and all the many variants produced by Maxis during the 1990s that used the same design template—did, however, have game-like *rules* that offered enjoyable play. Marketed as 'software toys,' the Sim designs avoided valuing game play as activity that necessitated winning or losing; the Sim software combined a game-like, rules-determined context—an embedded *context of control*—with more toy-like, self-directed, and self-expressive play. Yet, significantly, much of the appeal of the Sim series of games was a consequence of their faux reality. Insofar as the Sim games were marketed and valued as 'real,' they then had real-world consequences and significance—including commercially appealing educational applications.

Without directly employing Wright's innovative spreadsheet-based design template, current digital 'sandbox' games offer a toy-game hybrid play experience based on play within 'open-ended' virtual environments. *Microsoft Flight Simulator* (Microsoft 1977), which predates both *Tamagotchi* and *SimCity*, well represents the persistent appeal of this type of hyphenated simulation-game, a category of commercial game software that has grown into its own, broad, and best-selling market genre.

Open-ended play (which is, of course, only 'open-ended' in the fixed context of the game software) is now featured prominently in many of the digital game industry's most successful games and series—e.g. *Grand Theft Auto* (Rockstar 1997), *Assassin's Creed* (Ubisoft 2007), *Skyrim* (Bethesday 2011) and a wide variety of MMOs. In some cases, more expressive, toy-like play in these sandbox games is only available between segments of traditional game play. In other cases—e.g., the LEGO-like *Minecraft* (Mojang 2011)—toy-like play becomes the norm fairly quickly after preliminary and perfunctory traditional game play that introduces the player to the game's embedded rules that thereafter constitute its 'sandbox.'

A sandbox is a natural affordance for human play, offering human physical manipulators (hands) and a manipulable physical object (sand). Digital media can likewise serve as a natural affordance for human play with non-physical, *semiotic* objects. However, should the sand in the physical sandbox become fixed, more solid rock than granular sand, then these natural affordances for play are lost. Likewise, should the semiotic properties of digital-based signifiers become fixed in their references to non-digital signifiers, then semiotic affordances for play are lost.

Traditional games are unique configurations of signifiers that make appealing use of the pliable and polysemic nature of the signification process, including that process that signifies *self*. Digital media are capable of extending the semiotic properties of traditional games, yet they also seem capable of diverting them into the equally appealing though not quite so profound pleasures of ego gratification. This has already occurred with digital games configured as toys, and it seems to occur most seductively when those digital games configured as toys are also configured (and marketed) as toys that are *real*.

Conclusion

Digital media have demonstrated a particular affinity with games, making digital games more popular and more accessible than their traditional counterparts. However, this apparent affinity—based on observations of digital game sales—is somewhat misleading in terms of how digital media enforce the unique semiotic properties of games.

For instance, despite multiple versions of digital chess games, including versions that play the game better than any human player, digital media have failed to popularize chess and chess playing as much in the United States as Bobby Fischer did in a predigital age more than 40 years ago. Indeed, upon reflection, it may be that digital media have been most successful in promoting game sales by adapting the semiotics of toys to traditional game forms.

These newly popular digital games then become increasingly popular as *simulations* of toys, adopting the twin and mutually reinforcing guises of toy-as-cultural-artefact and toy-as-real. Digital media display a 'particular' affinity with games precisely because the rules of the game are more subject to digital manipulation than are the materialities of the toy: i.e. the materialities of the toy can be experienced as a fixed (code-determined) *context of control* in a digital game.

If so, then it is the pliability of the peculiar and paradoxical signifiers of traditional games that, somewhat paradoxically, has led to their conceptualization and use as something other than traditional games in digital media contexts: the semiotics of the game has proven both more controlling and more controllable than the semiotics of the toy.

References

Best, J. (1998). Too much fun: Toys as social problems and the interpretation of culture. *Symbolic Interaction, 21*(2), 197–212.

Bloch, L., & Lemish, D. (1999). Disposable love: The rise and fall of a virtual pet. *New Media Society, 1,* 283–303.

Copier, M. (2005, June). *Connecting worlds. Fantasy role-playing games, ritual acts and the magic circle.* Paper presented at DIGRA Conference, *Changing Views: Worlds at Play,* Vancouver. Retrieved June 1, 2014, from http://www.digra.org/digital-library/publications/connecting-worlds-fantasy-role-playing-games-ritual-acts-and-the-magic-circle.

Cr1TiKaL. (2011, April 9). *Facade game play and commentary.* [Video file]. Retrieved June 1, 2014, from http://www.youtube.com/watch?v=jvQB223IwYU.

ECSIP Consortium. (2013). *Study on the competitiveness of the toy industry.* Netherlands: ECORYS. Retrieved June 1, 2014, from http://ec.europa.eu/enterprise/sectors/toys/files/reports-and-studies/final-report-competitiveness-toys-ecsip_en.pdf.

Fleming, D. (1996). *Powerplay: Toys as popular culture.* Manchester: Manchester University Press.

Higuchi, T., & Troutt, M. D. (2004). Dynamic simulation of the supply chain for a short life cycle product—Lessons from the Tamagotchi case. *Computers & Operations Research, 31*(7), 1097–1114.

Huizinga, J. (1955). *Homo ludens: A study of the play-element in culture.* Boston: Beacon Press.

Kyburz, J. A. (1994). "Omocha": Things to play (or not to play) with. *Asian Folklore Studies, 53*(1), 1–28.

Leja, M. (2000). Peirce, visuality, and art. *Representations, 72,* 97–122.

Levaniouk, O. (2007). The toys of Dionysos. *Harvard Studies in Classical Philology, 103,* 165–202.

Malaby, T. M. (2007). Beyond play: A new approach to games. *Games and Culture, 2*(2), 95–113.

Mateas, M., & Stern, A. (2003). Façade: An experiment in building a fully-realized interactive drama. Paper presented at *Game Developer's Conference: Game Design Track*, San Jose, CA. Retrieved June 1, 2014, from http://lmc.gatech.edu/~mateas/publications/MateasSternGDC03.pdf.

Montessori, M. (1967). *The absorbent mind.* New York: Delta.

Pennell, G. E. (1994). Babes in toyland: Learning an ideology of gender. In C. T. Allen & D. R. John (Eds.). *Advances in Consumer Research* (Vol. 21, pp. 359–364). Provo, UT: Association for Consumer Research.

Piaget, J. (1962). *Play, dreams, and imitation in childhood.* New York: Norton.

Suits, B. (2005). *The Grasshopper: Games, life and Utopia.* Ontario: Broadview Press (Original work published 1978).

Sutton-Smith, B. (1984). A toy semiotics. *Children's Environmental Quarterly, 1*(1), 19–21.

Sutton-Smith, B. (1986). *Toy as culture.* New York: Gardner Press.

The Strong. (2013). *National toy hall of fame: Inducted toys.* Retrieved December 1, 2014, from http://www.toyhalloffame.org/toys.

Verenikina, I., Harris, P., & Lysaght, P. (2003, July). *Child's play: Computer games, theories of play and children's development.* Paper presented at IFIP Working Group 3.5 Conference: *Young Children and Learning Technologies.* University of Western Sydney Parramatta, Sydney, Australia. Retrieved June 1, 2014, from http://crpit.com/confpapers/CRPITV34Verenikina.pdf.

Vygotsky, L. (1933). Play and its role in the mental development of the child. In J. Bruner, A. Jolly, & K. Sylva (Eds.). *Play: Its role in development and evolution* (pp. 461–463). New York: Penguin.

Williams, D. (2006). A (brief) social history of video games. In P. Vorderer & J. Bryant (Eds.). *Playing computer games: Motives, responses, and consequences.* Mahwah, NJ: Lawrence Erlbaum. Retrieved June 1, 2014, from http://dmitriwilliams.com/WilliamsSocHist.doc.

Zimmerman, A. (2009, October 22). Mattel hopes *Barbie* facelift will show up younger rivals. *Wall Street Journal.* Retrieved June 1, 2014, from http://online.wsj.com/news/articles/SB125607851547797455.

Author Biography

David Myers is Distinguished Professor of Communication, School of Mass Communication at Loyola University, New Orleans, USA.

PART II

Language Development

CHAPTER 5

Age Differences in the Use of Toys as Communication Tools

Amanda Gummer

Introduction

Communication is a complex skill that requires learning to understand and speak a language, listening and responding appropriately to others, and learning to externalize thoughts and feelings (both verbally and non-verbally). Toys and games can support the development of communication as well as being used by children as tools with which to communicate specific pieces of information. For the purpose of this discussion, 'toy' is defined as any object that has been produced or manufactured to be played with, usually by a child. It does not include products that children may play with when the primary purpose of the object is not to support, facilitate or encourage playful behavior.

Toys facilitate communication opportunities in a number of ways, including:

- Role Play, for example with a toy kitchen; children can explore their identities by trying out different personalities and seeing what kind of

A. Gummer (✉)
Hertfordshire, UK
e-mail: amanda.gummer@fundamentallychildren.com

L. Magalhães and J. Goldstein (eds.), *Toys and Communication*,
https://doi.org/10.1057/978-1-137-59136-4_5

reaction they get. They can also express their identities through other creative activities, such as art and music.

- Competing, such as in a board game; they might be negotiating, discussing strategy, and more.
- Expressing their feelings. Children might find it difficult to vocalize their concerns, but a lot of anxieties can come out when they are playing in their fantasy worlds, acting as a character, or talking through puppets.

As they develop, children benefit from different types of play. Toy use evolves throughout childhood. They start being used, mainly by adults, as tools that improve a child's communication skills. Their use then evolves to make them tools that are used by children to facilitate communication both with peers and adults, and then are used as vehicles through which children are able to communicate aspects of themselves (including opinions and sense of identity). This chapter discusses how children of different ages tend to use toys for different communicative purposes. Evidence for this report is based on feedback obtained from watching children aged 0–12 play with over 800 toys as part of the Good Toy Guide evaluation process. Each toy was tested on at least four occasions with different children, and observations on the type of play, interactions, and communications have been recorded and analyzed for this chapter.

0–2 Years: Shared Attention, Babbling and Imitation

A baby's first attempts at communication start from birth, and involve crying (different cries will indicate different needs), gurgling, babbling, squealing, laughing, and smiling. Whilst babies do not actively use toys to communicate during the first few months, the presence of toys makes adults more playful and therefore promotes social interaction and non-verbal communication with even a young baby. At this age, whilst the toys are not being used as conscious tools for communication by the child, they are facilitating communication in important ways that warrant discussion here.

Music is possibly the most important form of non-human communication for a very young baby. The well documented 'Mozart effect' has been shown to have many developmental benefits for children, including some linked to communication, e.g. language learning. It is worth noting here that research by Hughes (1998) has demonstrated that it is not just classical music that produces the Mozart effect but any music with a rhythm. Therefore, toys that produce rhythmic tunes can also be argued to facilitate communication.

As a baby develops, he or she will also begin to take part in an early form of conversation—he or she might babble, then if the caregiver responds to this, the baby may react with a smile, in turn eliciting encouragement from the caregiver. This turn-taking shows awareness of one another and develops an understanding of the back-and-forth pattern of social interaction. This communication can be enhanced and encouraged by certain toys. For example, brightly patterned toys and particularly those that make sounds (such as a rattle) can attract the baby's attention and encourage a response. Adults with access to 'baby' toys—in particular, puppets—increased adult interaction with the babies and therefore promoted increased communicative experiences for the baby.

As babies learn to sit up and make more deliberate actions, they can also learn about the reciprocity of communication with games, such as rolling a ball between the baby and the caregiver, or playing peek-a-boo by hiding behind a toy, their hands, or any other prop.

From around 6 months old, babies will attempt to imitate adult speech and other familiar sounds through babbling. For instance, an adult might say 'teddy' every time they interact with a particular soft toy and the baby may try to copy the word without necessarily pronouncing it correctly. When the sound is sufficiently distinct to identify the object (e.g. teddy), parents tend to be very ready to understand the babble-speak term for the item, often adopting it themselves when referring to it; thus the toy acts as a tool to facilitate some of the earliest reciprocal verbal communication between an infant and their caregiver.

Babies may also use appropriate sounds with certain toys, showing that their understanding links to that object. If playing with a toy car, for example, a baby might make a 'brrmm' noise.

By 2 years old, a baby's language will grow into recognizable labels for specific objects and people, and he or she may be using one- or two-word sentences to convey meaning.

Shared attention and pointing are another part of early communication. By around 12 months, babies will be able to point to or reach for objects to show that they want them. Placing a favorite toy out of reach can motivate babies to communicate their wish to have the toy.

Caregivers can try to attract the baby's attention by looking at or pointing to a toy or object as well. Babies will also be able to point to a toy or object when asked, demonstrating their understanding of language and intention (even if they can't speak yet).

As well as showing their wishes and intentions, a baby will begin to show his or her own emotions—for example, by laughing to show happiness. Interacting with a baby through play can prompt these expressions and reinforce their communicative purpose, thus increasing the baby's communication skills.

2–4 Years: Language Explosion, Following Instructions and Pretend Play

Between 2 and 3 years old, a child's vocabulary typically expands from around 50 to over 1000 words. Verbal communication is improving as children start to talk in three- to four-word sentences.

The ability to understand how a word or sound can depict an object may be associated with their knowledge of symbolic representation, which has a strong connection with pretend play (for example, imagining that a kitchen is real or that a soft toy is alive). There are many toys that promote and encourage this type of play. Toys such as Playmobil's 123 range can both extend vocabulary (e.g. naming animals) and increase opportunities for imaginative communication narratives. The bright colours and simple construction activities appeal to children and increase the opportunities to engage in this type of play.

Listening is an important part of children's communication skills, and toys that encourage children to listen, such as musical toys, interactive books that are read by an adult, or interactive plush toys involving stories, can help them practice this. Musical toys are beneficial in helping children develop more communication skills than just listening. Toy instruments can be used to teach turn-taking and reciprocity—for example, the adult playing a sound first and the child responding.

This is also the age when children are able to follow simple directions, and toys can be used to encourage this. For example, a child might be asked to pick up a certain toy and put it in the toy box. Similarly, they might start playing games that require simple instructions to be followed. Children were observed following instructions and also asking simple questions or gesturing to clarify instructions when they didn't understand.

Once children can communicate verbally and have developed the ability to ask for clarification, they build on their knowledge of the world with 'how,' 'what,' and 'why' questions, and toys are often the prompts for these inquiries. Questions in the observation sessions were common with verbally competent 2-year-olds. For example:

'Can teddy drive the car?'
'Where does (sic) these (bricks) go?'
'When my turn? It my turn now?'

Sensory play gives children a chance to use descriptive words while verbalizing their experiences, which is particularly successful when an adult or older child encourages the child's conversation—for example, by saying a drum is loud or quiet, or asking the child whether the Play-Doh is soft or hard. The child's explanation may be accompanied by gestures and the social interactions obtained from the other people in the child's environment were observed to have a big impact on how willing children seemed to volunteer verbal descriptions. This suggests that, whilst toys are props for communication, the key factor in children's developing communicative competence is the interaction between a child and other people in his or her environment.

Many of the 2-year-old children observed played primarily in parallel and did not voluntarily interact with other children unless a child had taken a toy from them or had a toy that the first child wanted. However, many children were observed playing alone but muttering as they narrated their own stories and had pseudo-conversations between the characters in their fantasy worlds, thus reinforcing verbal communication through sub-vocal repetition.

As children begin to play socially with each other (usually from 3 years), their communication skills seem to explode and skills such as negotiation are employed. The toys that children choose at this age tend to reflect their enjoyment of communication, and we observed children playing with a Play School role-play set in a way that promoted cooperation and collaboration with peers, as well as providing a tool for the children to explore and overcome their preconceptions, expectations and anxieties about starting school.

We observed many different toys that prompted and supported imaginative social play. As well as speaking within their roles, the children discussed the plot and negotiated with the other children (for example, deciding who can cook or who gets to play with the special blue car). Children also showed a surprisingly advanced ability to empathize and understand others' thoughts and feelings whilst playing with imaginative role-play toys, particularly when the themes were personally relevant to the child. Childcare workers and parents reported that the children weren't usually as willing or able to empathize outside of these imaginative play scenarios.

Imaginative play and the toys that support it are also beneficial in promoting communication between children and their caregivers. Through a child's fantasy worlds and stories, he or she can communicate in a way that might not be possible in a direct conversation with an adult. Children may act out scenarios that worry them, helping them to explore their emotions and also giving an overseeing adult the opportunity to support this by asking questions of the character, rather than of the child. This indirect communication provides an important shield for the child and encourages him or her to share feelings more openly, as he or she feels less exposed or vulnerable to having those feelings misunderstood or not validated.

Preschool children are increasingly using tech-based toys to communicate. Much of this play is similar in character and play patterns to non-tech play and much of it involves communication. Tech products that combine digital with physical play have been successful in promoting story-telling and decision-making abilities in young children. Many tech toys have addressed the criticisms of solitary, passive sedentary play, and there are now many preschool tech-based toys and games that both promote communication skills and assist children with their communication. Play workers remarked on the increasing popularity and availability of interactive plush toys. Some children were observed talking to an interactive bear in a more animated manner that was closer to natural communication (as observed between friends and between the children and their carers) than when playing with an inanimate plush toy, although others communicated in a similar manner regardless of whether the toy was interactive or not. For the children who differentiated their communication style depending on the type of toys, we observed that communication with favorite plush toys was more of a 'confidante' nature and some children tried to hide the fact that they talked to their plush toys whereas, with the interactive toys, children were very open about talking out loud to the toy and involved it in more playful behavior patterns.

4–7 Years: Self-expression, Identity and Belonging

Children continue to enjoy pretend play as they progress through the first years of school. Toys that facilitate imaginative play, where they can explore anxieties and express themselves through made-up characters and stories, add a richness to their dialogues and prompt explorations of play themes and behaviors that may not have occurred without the toys. What is not certain and would be worthy of investigation is how children's imaginative play is impacted by the presence or absence of toys that facilitate imaginative play. It would be interesting to examine whether children play for longer or shorter times and whether the toys limit or expand children's play patterns and whether, if any effect is observed, there is any gradation within that effect based on toy type or design. There is currently debate over whether licensed toys promote or hinder imaginative play, and further research is needed to properly inform this issue.

Pretend play can be used by adults at this age to encourage discussion around certain topics. For example, a toy farm might be used to prompt a discussion on where our food comes from and is often used in play therapy. The observations made for the purpose of this research did not include therapeutic settings, although play workers did use toys as a distraction for children who were, at times, upset or alone.

As children get older they move toward more planned 'performance'-type play, acting or using puppets to act out the scenarios. There is a wide range of products that support and encourage children to act and perform. Observations of children playing with a pop-up theater illustrated how children will often benefit from adult support to begin a performance, but quickly take ownership of the dialogue and plot and develop the play along their own, socially negotiated path.

Toys that children can relate to seem to promote self-expressive and explorative communication more than toys within a fantasy-themed design. For example, life-like dolls were observed to prompt more self-referencing imaginative play and involve narratives in which the child was the main protagonist in a way that wasn't observed in children playing with less realistic dolls or plush toys.

Children also start to discover or form their identities, and one of the most accessible ways of communicating this developing sense of identity, preferences, and values, is through the toys they play with. Toys that have flexibility for personalization and facilitate children's expression of exaggerated traits or personalities seem very popular with children of this age, which suggests that children benefit from feeling more able to express their identity through play, using the toys, and it would be interesting to research whether a child's development is negatively impacted if they have not been given access to these types of toy in early childhood.

The play patterns that are employed to express this developing identity can vary. Children may try out different characters through pretend play—for instance, if a child is shy, they might pretend to be a strong, confident superhero, or a child growing up in a sporty family may role-play being a musician. Children might also use art, music and dancing to express their developing personalities. Toys that encourage dressing up can help children explore different personas and communicate personal preferences in a way that children may find difficult without such toys.

Team games become more popular as children move towards the upper end of this age band; they give children the opportunity to develop their emerging collaborative and strategic abilities to work with others towards a shared goal. This also reflects the increasing importance of belonging to a group, as children begin to notice the similarities and differences between one another. Toys that are aligned to a particular club or brand seem to be embraced by children, especially boys, as part of their developing identity. Football clubs and toys or games that allow them to demonstrate their allegiance to a particular club were noted as being particularly appealing to boys at this age.

7–12 Years: Self-awareness, Complex Thought/Conversation and Expressive Writing

Identity and belonging are particularly important as children reach the end of their primary school years. Children often communicate their identity through their chosen hobbies, and these may be influenced by a desire to rebel against parental expectations, along with a (sometimes conflicting) desire to be accepted by their peers.

Children seem to move away from specifically designed toys and pretend play as they get older, but they still display a lot of playful behavior and elaborate play patterns—much of this with the aid of props but not items that children themselves refer to as 'toys.'

Play that develops communication revolves more around board games and sports as well as tech-based social interactions. Group games give children the chance to practice taking turns, and learn about friendly competition and consolidate the child's sense of belonging which feeds into his or her sense of identity.

Self-expression through the arts—drawing, making, painting, music, and performance—becomes more complex and personal, with the final product a child's main focus of pride. The styles they choose to include show the child's identity; for example, they might use big and loud movements when performing, or bright bold colors in their artwork. The fact that they have made the decision to use these in their pieces communicates their personality and feelings when creating it.

Children's cognitive abilities develop quickly at this age, and exploration and investigation are popular. The rise in the availability and range of science-based toys and activities is testament to how hungry children are for knowledge and to build on their basic understanding of how the world works through scientific activities, such as experimentation and observation. Nature sets that include creating habitats for bugs and science kits that involve carrying out experiments across all science disciplines were observed as being very popular with both girls and boys in this age group. With some products, there were design flaws that reduced the accessibility or enjoyment of the activity itself, but the enthusiasm of the children for engaging with the activity was high.

As their vocabulary builds, children might also experiment with a more expressive writing style and will now be capable of exploring made-up characters and fantasy worlds through their writing. The type of world they set their stories in, and the way the characters interact, can express the thoughts and anxieties of the child. There are toys and games that help children develop their imaginative story-telling skills. Story Cubes appeal both to children who enjoy writing stories and to those who prefer telling them verbally in a more spontaneous manner. Historical play-sets give children elaborate contexts in which to immerse themselves and create their own interpretation of different eras. This playful creation of virtual worlds is also observed when playing digitally with some apps and online games. Children were observed playing in very similar ways both on- and off-screen when creating and engaging in make-believe worlds. If children use digital toys in much the same way as they use physical toys to explore their environments and create new ones, this raises the question: does the social aspect of many digital toys make them a better mode for communication than their physical counterparts?

Conclusion

As children develop, their use of toys evolves: from tools used by adults to support interaction through objects that promote communicative abilities in young children to instruments of communication for children to explore and express their identities, values, and preferences.

Early communication is through simple gestures, turn-taking, and shared attention. The importance of toys in this stage is to gain a baby's interest and facilitate communicative interaction with caregivers.

A baby's first foray into vocalization involves babbling and making familiar sounds, and these are often related to the toy they are playing with. As children learn to talk, it is the opportunity for social interaction which supports their development of communication skills, but children do not seem to deliberately use toys to communicate until they are at least 2.

Pretend play allows children to converse around their play, negotiating and empathizing with others, and is particularly popular with 3–7-year-olds. At this stage, children are starting to use toys as props to express their developing sense of identity as well as tools for facilitating conversation and communication with others.

As children become older, board games and sports become important types of play for encouraging social interaction as well as ever-expanding digital play opportunities.

Children also learn to express themselves and their identities through play. From a young age, children do so by acting as different characters, in pretend and small-world play when they are younger and through performance and stories as they get older. From the scenarios and characters children create, adults can understand more about problems and worries the child is having that they might not be able to discuss in direct conversation.

As children become competent writers, the fantasy worlds they used to explore through pretend play can transfer from the physical world to pen and paper and digital worlds. Here the stories become more about the child reflecting on and communicating their thoughts, feelings and anxieties to the reader or other members of their digital environment. The social interactions and mutually rewarding friendships are not diminished at this stage; rather, they are enhanced by the children's ability to use language to express themselves and reflect on their experiences and emotions.

Toys contribute to children's communication in a variety of ways and, apart from passive solitary play, all toys can be used as tools to help children communicate—whether that is learning to speak, write, and act or to aid the communication of particular messages, toys are important props in children's social and communicative development. From birth, children can learn from the social interactions that play encourages, while imaginative play in particular (both creative and pretend play) helps them communicate thoughts and feelings through different means.

References

Alexander, G., & Hines, M. (2002). Sex differences in response to children's toys in nonhuman primates (cercopithecus aethiops sabaeus). *Evolution and Human Behaviour*. http://www.ehbonline.org/article/S1090-5138(02)00107-1/abstract.

David, A. (2014). *Help your child love reading*. London: Egmont.

Foster, S. (1990). *The communicative competence of young children*. London: Longman.

Gummer, A. (2015). *Play*. London: Vermilion.

Hirsh-Pasek, K., & Golinkoff, R. M. (2003). *Einstein never used flash cards*. Emmaus: Rodale.

Hughes, J. (1998). The "Mozart effect" on epileptiform activity. *Perceptual and Motor Skills, 86*, 835.

Mortimer, H. (2008). *Making the early years foundation stage work for you (30–60+ months)*. Stafford: QEd Publications.

Rowlands, H., & Mortimer, H. (2008). *Making the early years foundation stage work for you (0–36 months)*. Stafford: QEd Publications.

Sunderland, M. (2006). *The science of parenting*. London: Dorling Kindersley.

Snowling, M. (1987). *Dyslexia: A cognitive developmental perspective*. Oxford, UK: Blackwell.

Sutton-Smith, B. (1997). *The ambiguity of play*. Cambridge, MA: Harvard University Press.

Wood, E., & Attfield, J. (2005). *Play, learning and the early childhood curriculum*. Thousand Oaks, CA: Sage.

Author Biography

Dr. Amanda Gummer is founder of Fundamentally Children and Research Director for the social enterprise Families in Focus, co-founder of Karisma Kidz, co-chair of the UK Chapter of Women In Toys, member of the British Psychological Society, and of the International Toy Research Association.

CHAPTER 6

LMNOBeasts™: Using Typographically Inspired Toys to Aid Development of Language and Communication Skills in Early Childhood

Todd Maggio, Kerri Phillips and Christina Madix

Review of Literature/Support

Teaching children sound to symbol matching is considered necessary in learning a language. Learning that a sound can be represented by a written symbol is a precursor to literacy development. It is thought that an effective communicator should become a literate person due to the underpinnings of this early developing relationship in childhood.

T. Maggio (✉) · K. Phillips · C. Madix
Louisiana Tech University, Ruston, USA
e-mail: maggio@latech.edu

K. Phillips
e-mail: kphillip@latech.edu

C. Madix
e-mail: cmadix@latech.edu

L. Magalhães and J. Goldstein (eds.), *Toys and Communication*,
https://doi.org/10.1057/978-1-137-59136-4_6

Development Basis

Joint attention is a benchmark for developing theory of mind and narrative skills during children's preschool years. Joint attention is defined as an engaged interaction between a child and caregiver with shared objects or events (Adamson et al. 2004; Mendive et al. 2013). The mother-infant-object triadic interaction is often the focus of research in early communication skills. The importance of this interaction has been studied as the development of sharing attention, following attention, and directing attention, and is believed to lead to the development of a language (Beuker et al. 2013; Tomasello and Farrar 1986). The development of attention in language development leads to the intentionality of communication. In other words, intention is considered important in joint attention.

The intent to communicate means that two individuals develop a shared social interaction, which can be directed by either person. The development of joint attention is necessary in order for a child to become an active communicator and direct the social interaction. Otherwise, the child will become a passive communicator and only respond to the social interaction directed by others.

In terms of age of development, the triadic interaction can emerge as early as 3 months but most often in 6-month-olds and continues to develop until 12 months of age (Striano et al. 2006). At 12 months, infants begin to use language and interpret concrete information independently. The caregiver remains crucial to the developmental process until 18 months of age for interpretation of more complex information. By age 2 years, the child begins to independently interpret complex information (Beuker et al. 2013; Mendive et al. 2013).

The development of this triadic interaction leads to the ability to regulate one's own behavior as well as the behavior of others. Studies have indicated that, between 9 and 18 months of development, children begin to respond to joint attention (Van Hecke et al. 2012). Van Hecke et al. stated that, by 12 months of age, the development of joint attention is crucial in the later development of self-regulation at 36 months of age.

In terms of theory of mind, the development of joint attention is essential. Theory of mind is a concept that people have points of view, beliefs, and/or emotions that allow successful communication with others. Individuals with communication deficits may have difficulty understanding that others have different perspectives, worldviews, and/or emotions other than their own, which results in deficient social interactions. In some instances, children with communication delays or disorders are thought to not have theory of mind. These children may lack the development of joint attention necessary in establishing meaningful communication with others. For instance, they may lack pragmatic skills necessary for social communication purposes. Additionally, children with autism are thought to lack presupposition when communicating. Presupposition requires the speaker and listener to be aware of the communicative intent (Paul and Norbury 2012). In the development of communicative intent, specifically in presupposition, the use of symbols and/or toys is often used to develop joint attention in the triadic dyad.

Research has shown the use of symbols during caregiver interaction helps establish joint attention for communication purposes (Adamson et al. 2004; Newland et al. 2001). Young children were found to attend best to structures that were distinctly asymmetrical (Vallotton and Ayoub 2010). Children between 6 and 18 months who participated in caregiver-directed joint attention with symbols were socially able to initiate joint attention with a communication partner by the time they were 18 months of age. In early development, the use of symbolic play allows children to develop narrative skills by organizing, sequencing, and establishing a main idea, and understanding the perspective of a communication partner (Adamson et al. 2004). Typically, the ability to initiate and coordinate joint attention develops around 30 months of age. The amount of time that children between ages 18 and 30 months engaged in joint attention tasks using symbols greatly influenced expected receptive and expressive language skills measured through assessment. Adamson, Bakeman, and Deckner proposed that research should be conducted beyond the second year to better understand how children organize symbols for use in reading, drawing, writing, and engagement in discussions about past and future events. Language skills during preschool years can predict academic success once children enter kindergarten.

Numerous approaches currently exist for providing young children with instruction in sound to symbol matching (Fisher 2008; Gelzheiser 1991; Hulme et al. 2012; Woods et al. 2005). For example, Seward et al. (2014) investigated the use of a digital product in teaching reading to children considered to be at risk. Gelzheiser (1991) stated that the use of nonsense words with unfamiliar sound/symbol letters may generalize better when decoding than to use known letters. Woods et al. (2005) indicated that younger children perform better in reading tasks with the use of larger font size specifically using the sans serif font of Arial.

One of the areas regaining attention is actually related to an older debate on the use of typography in teaching a child to recognize symbolic materials. Early studies were conducted on the use of 2-D and 3-D letter forms to aid in the discrimination of 2-D letters (Thornburg and Fisher 1970; Towner and Evans 1974). A 2-D letter form is a form that has dimensions of height and width. A 3-D letter form has height, width, and depth.

Thornburg and Fisher (1970) reported in an exploratory study that children of 3–4 years of age may discriminate 3-D forms more accurately than 2-D forms. The 3-D forms allowed a more complex representation of the visual schema, thus allowing fewer discrimination errors and allowing for greater generalization of 2-D forms in a variety of shapes, including print. Towner and Evans (1974) also stated that 3-D letter forms aid in generalization of letter recognition and discrimination faster than using only 2-D materials.

Woods et al. (2005) investigated typeface and font size in tachistoscopically presented letter pairs in children (kindergarten through 4th grade). Young children (K–1st grade) performed better when presented with larger font. The investigators attributed this to the font being more legible for the K–1st grade as opposed to the 2nd through 4th grade, due to memory and attention.

The studies presented have focused on children without known communication deficits who are in the process of learning a sound to symbol system. The proposed research continues to focus on children without known communication deficits and is designed to be practical and cost-effective in terms of human and material resources required. The goal is not to determine the superiority of a particular supplemental method of sound to symbol matching; rather, it is to determine if an abstract independent learning activity is at least as effective as a traditional-based intervention.

However, the needs of all children may be served better if teachers have access to supplemental materials/activities for instruction. Children at risk for failure to learn sound to symbol correspondence as presented through the curriculum need additional support. At-risk children tend to need additional modalities in order to develop sound to symbol correspondence.

As technology continues to shape the interaction of children with images and the growing concern of literacy, educators should employ more types of multisensory images. In other words, only using 2-D images to learn the sound to symbol correspondence may not be the most effective strategy.

By using symbols that are strongly influenced by shaping graphic print, there is potential for the symbols to influence visual recognition of letters for literacy and language during toddler and preschool years. Using caregiver interaction and joint attention with the graphic print symbols, this project will investigate whether typically developing children in preschool classrooms can receptively identify typographic symbols that are incorporated into typo-zoomorphic compositions.

Pilot Study

Purpose

The purpose of the pilot study was to investigate whether children aged 3 through 5 years old could recognize symbolic letters embedded within the LMNOBeasts™ prototypes. Data collected targeted identification and matching LMNOBeasts™ prototypes to the five symbolic letters presented each week.

Method

The pilot study was a quasi-experimental design. Children aged 3–5 years were recruited from two local preschools (one a local Montessori center and one a traditional preschool center). All the children continued to receive normal instruction in their classroom settings. Pre-and post-testing was completed for the children before and after the 3-week exposure. The children who participated were shown five 2-D symbolic letters and a 3-D LMNOBeasts™ prototype each week for 3 weeks. Data collected revealed the children's abilities to identify symbolic letters and match the letters to the LMNOBeasts™ prototypes.

Participants

A total of 43 students were recruited for the pilot study. Twenty-four children were from the Montessori center. Nineteen students were from the preschool center. After completing the pre-test, 38 children met all criteria of participation for the pilot study. Criteria for inclusion was age, no visual, auditory, cognitive, and/or physical disability or disorder that would limit the use of hands, English as primary language, does not have an Individualized Education Plan (IEP), non-reader, and received a score of less than 80% correct on the researcher-designed pre-test. The criterion of 80% correct was based on a ten-point grading scale in the local school district. This percentile ranking is equivalent to a letter grade of B.

Adult volunteers consisted of university students who provided written consent to participate in the project. Volunteers were recruited from Louisiana Tech University classes after hearing a brief description of the research project provided by one of the researchers. A total of four volunteers served as data collectors for the project. Each volunteer received training focused on presentation of the materials and how to mark answers on the data sheet. Student volunteers did not receive course credit or other compensation for their volunteer work.

Materials

Customized Pre- and Post-test

Researcher-developed pre- and post-tests in the form of matching alphabetic symbols were used to document the children's progress. The pre- and post-tests contained questions about prior symbolic letter knowledge the participants may have had before the exposure. The pre-and post-tests were black and white simple symbolic letter representations of the traditional English alphabet.

Symbolic Letters

Three sets of five symbolic letters were printed and laminated for use during the weeks of exposure. The three sets were influenced by the pedagogical system of Montessori. Montessori divides symbolic letters into sets of five (four consonants and one vowel in each set). The sets utilized in the pilot study were set 1 (a, b, s, m, t), set 2 (e, l, f, k, d), and set 3 (i, c, r, h, g).

LMNOBeasts™

Researcher-developed prototypes, LMNOBeasts™, were used in the study. The LMNOBeasts™ prototypes were designed as 3-D typo-zoomorphic forms composed from five symbols based on a pedagogical Montessori influence. Conceptually, the LMNOBeasts™ models are typo-zoomorphic compositions using segments of letter forms to articulate animal anatomy.

This approach is based in part on research that shows segmental analysis as more effective for learning to perceive letters than dynamic tracing (Courrieu and DeFalco 1989; Towner and Evans 1974). The chosen letters, combinations, and physical appearance are based on the Montessori method of teaching the sound to symbol correspondence. In the Montessori method, multisensory learning is a key component in learning letter forms. Letter forms are made using sandpaper type of letters to engage the student in multisensory learning. Thus, the physical test models were produced on a 3-D printer using standard PLA filament and finished with a blue sandpaper texture.

Data Collection
Volunteers collected data on a customized response sheet to indicate whether each target symbol was correctly identified and matched by the participants. This response method is a typical method used in pre-K class settings.

Setting
The settings for the entire project were a local preschool and a local Montessori school. The preschool serves approximately 22 students between the ages of 3–5 years old. The Montessori school serves ages 20 months–14 years old. Only the preschool classrooms, for ages 3–5 years old, were used in the pilot study.

The pre-tests, weekly tests, and post-tests took place in the regular education classroom. The weekly tests took approximately 5–10 min each week to administer. Volunteers called each participant individually to a table in the classroom. This is a typical place for teachers to provide instruction in the regular education classroom.

During the course of the exposure, typical interruptions occurred, such as school announcements, picture day, a field trip, special guests, and teacher training. Each type of interruption was handled differently, but did not compromise the schedule of testing or exposure, other than the day of the week or time of day that each participant was seen.

General Procedures and Setting

A meeting with teachers at both schools was held by the researchers prior to beginning the project. The project, including their role in the classroom, was explained to the teachers. Teachers were asked to continue classroom instruction as normal during the pilot study. Parent letters were sent home to the students who met the criteria of the project. Additional parent letters were sent home one more time for those parents who had not returned letters at the previous attempt.

Once parent letters were received, the students who were eligible to participate were given the researcher-designed pre-test. The pre-tests were scored by the researcher and those participants who scored below 80% participated in the exposure portion of the pilot study.

Training for Volunteers

The volunteers attended a training session prior to the beginning of the 3-week session. Training covered the background of the project, expectations of volunteers' responsibilities, and instructions for documenting their activities with each student. Basic information about procedures for volunteers in the preschool center and the Montessori center was provided, such as checking in and out in the main office and wearing identification provided by the school at all times.

General

Participants worked independently with a volunteer each week. The volunteer used a customized response sheet for each participant. The response sheet had directions and carrier phrases to guide the volunteers in how to introduce the new materials. This format was used as a guide for the volunteer in presenting information consistently between each volunteer and each participant. The volunteer placed each of the five symbolic letters in front of the child. The participant was asked to identify each of the five symbols. Data was collected either as 'yes' the participant did identify, or 'no' the participant did not identify, the symbolic letter correctly. The volunteer then presented the LMNOBeasts™ prototype to the participant by placing it on the table with the symbolic letters. The volunteer encouraged the participant to look at and feel the LMNOBeasts™ prototype. The participant was asked to find a part of the LMNOBeasts™ prototype that looked like each of the five symbolic letters. The volunteer recorded these responses on the customized response sheet. The participants completed the exposure tasks once a week for 3 weeks. An RCA VR5330R-B digital voice recorder was used during all interaction with the participants and volunteers. A week after the last exposure tasks the participants were given the post-test.

Fidelity in Treatment

The researchers observed each volunteer on two separate occasions to verify procedures were being completed as outlined in the methods. The observations served to document consistency among volunteers (e.g. time spent showing the symbolic letters and LMNOBeasts™ prototypes, how the questions were asked, and interaction of the participant). The customized response forms completed by the volunteers were used to support consistency of procedure. The digital voice recordings were used to verify consistency of exposure between volunteers and participants. The scoring of the pre-and post-tests, weekly customized response forms, and the entry of data into an Excel spreadsheet were validated by at least one volunteer and two of the three researchers.

Results

The purpose of the pilot study was to investigate whether children aged 3 through 5 years old could recognize symbolic letters embedded within the LMNOBeasts™ prototypes. This study employed a quasi-experimental design using a pre- and post-test to identify symbolic letter knowledge the participants had prior to and after the weekly exposure to symbolic letters and LMNOBeasts™ prototypes. All participants received typical classroom instruction during the pilot study.

Demographic Data

A total of 43 students were recruited for the pilot study. Twenty-four children were from the Montessori center. Nineteen students were from the preschool center. After completing the pre-test, 38 children met all criteria of participation for the pilot study. Once the pilot study was complete, the data of 14 participants was not used due to absences either during the weekly exposure or the post-test. There were 13 males and 11 females amongst the participants' data calculated in the pilot study.

Statistical Results

The pilot study was analyzed using paired *t*-tests and descriptive statistics. The dependent variable of time (pre-and post-test) was used to evaluate the two groups. The first analysis included all data from participants of both schools. Results indicate a significant difference between pre-test ($M = 7.74$, $SD = 2.79$) and post-test ($M = 6.5$, $SD = 2.65$), $t(23) = 2.09$, $p = 0.047$. Therefore, $p < 0.05$ and we can reject the null hypothesis [H_0: There is no difference between pre-tests and post-tests of the participants of the two schools (H_0:$\mho_{pr} = \mho_{po}$)].

Then data was broken into two sets, a pre-and post-test for each school. The participants at the early childhood center revealed results indicating a significant difference between pre-test ($M = 9.61$, $SD = 1.89$) and post-test ($M = 6.69$, $SD = 2.75$), $t(12) = 4.27$, $p = 0.001$. Therefore, $p < 0.05$ and we can reject the null hypothesis [H_0: There is no difference between pre-tests and post-tests of the participants at the early childhood center. (H_0:$\mho_{pr} = \mho_{po}$). The data from the Montessori center revealed results indicating a no significant difference between pre-test ($M = 5.54$, $SD = 1.92$) and post-test ($M = 6.27$, $SD = 2.64$), $t(10) = -1.14$, $p = 0.277$. Therefore, $p > 0.05$ and we cannot reject the null hypothesis [H_0: There is no difference between pre-tests and post-tests of the participants at the Montessori center (H_0:$\mho_{pr} = \mho_{po}$)].

Each weekly exposure was examined for growth in the areas of identification and matching abilities between the symbolic letter forms and the LMNOBeasts™ prototypes. Table 6.1 explains weekly means of the identification and matching of all the participants.

The data was broken into two sets, one for early childhood center and one for Montessori center. Table 6.2 shows results of the data collected during weekly exposures of identification and matching of the presented materials.

Although the data was not analyzed for a statistical difference, the results suggest that children could recognize the symbolic letters embedded in the LMNOBeasts™ prototypes. While identification of the letters did not demonstrate a strong understanding of the symbolic forms, the participants were able to match symbolic forms to the LMNOBeasts™.

Table 6.1 Weekly exposure average for all participants

Week of exposure	*Mean identification*	*Mean matching*
Week 1	2.0	2.33
Week 2	0.79	3.88
Week 3	0.96	3.08

Table 6.2 Weekly exposure average for participants by schools

Week of Exposure	*Mean identification*	*Mean matching*
Early childhood center		
Week 1	1.69	3.46
Week 2	0.69	3.92
Week 3	0.92	3.0
Montessori center		
Week 1	2.36	1.0
Week 2	0.91	3.82
Week 3	1.0	3.18

Conclusion/Discussion

Research continues to explore how children learn to code the orthographic representation of the sound system (Horbach et al. 2015; Whitehurst and Lonigan 1998). While the literature supports the use of developing new teaching strategies, one of the distinct differences with this project is the use of multisensory and/or multimodality learning experience with the typo-zoomorphic forms. The use of typo-zoomorphic forms may provide a unique way for children to store, process, and/or retrieve information in their short-term memory.

The purpose of the pilot study was to investigate whether children aged 3 through 5 years old could recognize symbolic letters embedded within the LMNOBeasts™ prototypes. Data collected targeted identification and matching LMNOBeasts™ prototypes to the five symbolic letters presented each week (set 1 (a, b, s, m, t), set 2 (e, l, f, k, d), and set 3 (i, c, r, h, g).

While the data is limited, the study did find that children could correctly identify 2-D letters within the 3-D typo-zoomorphic prototypes. It was interesting to note that children being instructed in the Montessori center were able make greater inferences when discussing the 3-D typo-zoomorphic forms. An area of investigation could be whether this difference can be attributed to the differences in the teaching pedagogy of the two centers (Lillard 2013). Montessori encourages the exploration of the environment through independent learning and imaginative play. The traditional preschool setting, on the other hand, encourages more teacher-directed learning. Additionally, the researchers used more of a teacher-directed style when interacting with the children from Montessori. The difference in pedagogical style may have influenced the findings between the two groups.

Further studies are needed before effectiveness of exposure can be made regarding the typo-zoomorphic prototypes on learning to recognize 2-D letter forms. The use of a control group to determine differences is needed before generalizations can be made. Also, an increase in sample size would assist in determining statistical significance, as would the development of a pre-test or post-test measure to examine perceptions of the LMNOBeasts™ in comparison to the traditional orthographic letters.

Another area of interest would be to investigate the differences between male and females in perceptions of the prototypes. Allowing children a longer exposure time to the LMNOBeasts™ as well as to the orthographic symbols may provide additional avenues of research.

The pilot study does provide a foundation for continued studies on the use of typo-zoomorphic shapes. While this study was small in sample size and length, there appears to be some evidence that the use of typo-zoomorphic symbols may support the notion that the use of simple segments in novel ways may aid the analytical processes needed for perceiving letters (Courrieu and De Flaco 1989; Woods et al. 2005). The value of new educational tools must be based on evidence that is grounded in learning theory and design theory in the quest to address problems in literacy.

Toy Design

The LMNOBeasts™ brand was launched with the goal of developing a typo-centric approach to designing toys that enhances the educational and therapeutic potential for beginning readers. This research lays the foundation for a more effective approach to adapting the formal language of type to the physicality of play. Studies have shown that children will increase their communicative attempts when engaged with novel objects. When presented with a new object, children tend to explore it through the use of gesturing, increasing vocalizations or attempts at labeling.

At the heart of LMNOBeasts™ is a co-creation model that leverages a child's innate play instinct to make them an active participant in the creative process. This typo-zoomorphic process is the true product as it informs the creation of a wide range of objects, primarily toys. Using a proposed mobile app and website, children can share their own creations with other 'Beasties', or see their designs come to life as custom-made toys. A multitude of possibilities exist thanks to new 'on-demand' manufacturing technologies—particularly 3-D printing which can deliver the experience directly to the classroom or home.

LMNOBeasts™ builds upon research into the significance of 3-D forms and segmental analysis to the process of letter recognition. By applying these ideas to the design of toys, children and their caregivers can achieve more productive gains in letter learning and communication skills.

References

Adamson, L. B., Bakeman, R., & Deckner, D. F. (2004). The development of symbol-infused joint engagement. *Child Development, 75*(4), 1171–1187.

Beuker, K. T., Rommelse, N., Donders, R., & Buitelaar, J. K. (2013). Development of early communication skills in the first two years of life. *Infant Behavior and Development, 36*, 71–83.

Courrieu, P., & De Falco, S. (1989). Segmental vs. dynamic analysis of letter shape by preschool children. *Cahiers de psychologie cognitive, 9*(2), 189–198.

Fisher, P. (2008). Learning about literacy: From theories to trends. *Teacher Librarian, 35*(3), 8–12.

Gelzheiser, L. M. (1991). Learning sound/symbol correspondences: Transfer effects of pattern detection and phonics instruction. *Applied Cognitive Psychology, 5*(4), 361–371.

Horbach, J., Scharke, W., Croll, J., Heim, S., & Gunther, T. (2015). Kindergarteners' performance in sound-symbol paradigm predicts early reading. *Journal of Experimental Child Psychology, 13*, 256–264.

Hulme, C., Bowyer-Crane, C., Carroll, J. M., Duff, F. J., & Snowling, M. J. (2012). The casual role of phoneme awareness and letter-sound knowledge in learning to read: Combining intervention studies with mediation analyses. *Psychological Science, 23*(6), 572–577. doi:10.1177/0956797611435921.

Lillard, A. S. (2013). Playful learning and Montessori education. *American Journal of Play, 5*(2), 157–186.

Mendive, S., Bornstein, M. H., & Sebastian, C. (2013). The role of maternal attention-directing strategies in 9 month old infants attaining joint engagement. *Infant Behavior and Development Journal, 36*, 115–123.

Newland, L. A., Roggman, L. A., & Boyce, L. K. (2001). The development of social play and language in infancy. *Infant Behavior and Development Journal, 24*, 1–25.

Paul, R., & Norbury, C. (2012). *Language disorders from infancy through adolescence* (4th ed.). St. Louis, MO: Elsevier.

Seward, R., O'Brien, B., Breit-Smith, A. D., & Meyer, B. (2014). Linking design principles with educational research theories to teach sound to symbol reading correspondence with multisensory type. *Visible Language, 48*(3), 87–108.

Striano, T., Chen, X., Cleveland, A., & Bradshaw, S. (2006). Joint attention social cues influence infant learning. *European Journal of Developmental Psychology, 3*(3), 289–299.

Thornburg, K., & Fisher, V. L. (1970). Discrimination of 2-D letters by children after play with 2-or 3-dimensional letter forms. *Perceptual and Motor Skills, 30,* 979–986.

Tomasello, M., & Farrar, M. J. (1986). Joint attention and early language. *Child Development, 57*(6), 1454–1463. doi:10.2307/1130423.

Towner, J. C., & Evans, H. M. (1974). The effect of three-dimensional stimuli versus two-dimensional stimuli on visual form discrimination. *Journal of Literacy Research, 6*(4), 395–402.

Vallotton, C. D., & Ayoub, C. C. (2010). Symbols build communication and thought: The role of gestures and words in the development of engagement skills and social-emotional concepts skills during toddlerhood. *Social Development, 19*(3), 601–626. doi:10.1111/j.1467-9507.2009.00549.x.

Van Hecke, A., Mundy, P., Block, J., Delgado, C., Parlade, M., Pomares, Y., et al. (2012). Infant responding to joint attention, executive processes, and self-regulation in preschool children. *Infant Behavior and Development, 35,* 303–311.

Whitehurst, G. J., & Lonigan, C. J. (1998). Child development and emergent literacy. *Child Development, 69*(3), 848–872.

Woods, R. J., Davis, K., & Scharff, L. F. V. (2005). Effects of typeface and font size on legibility for children. *American Journal of Psychological Research, 1*(1), 86–102.

Author Biographies

Todd Maggio, M.F.A is Associate Professor at the School of Design at Louisiana Tech University, USA.

Kerri Phillips, SLP.D is Professor and Program Director of the Graduate Program in Speech-Language Pathology at Louisiana Tech University, USA.

Christina Madix, M.A is an instructor in Speech-Language Pathology and Director of Speech and Language Services of the Louisiana Tech Speech and Hearing Center at Louisiana Tech University, USA.

PART III

Toys, Culture and Communication

CHAPTER 7

Images of Toys in Spanish Art (15th–19th Centuries): Iconographic Languages

Oriol Vaz-Romero Trueba and Michel Manson

To see a World in a grain of sand
and a heaven in a wild flower,
hold infinity in the palm of your hand,
and eternity in an hour.
WILLIAM BLAKE (*Auguries of Innocence*, 1863).

All illustrations in this chapter are original drawings by Oriol Vaz-Romero Trueba, in artistic collaboration with Esther Alsina Galofré. These images have been made with pen and black ink on paper in order to clarify the forms and objects of the original masterpieces referenced in each case.

O. Vaz-Romero Trueba (✉)
Universitat de Barcelona, Barcelona, Spain
e-mail: ovaztrueba@gmail.com

M. Manson
Université Paris-XIII, Paris, France
e-mail: michelj.manson@wanadoo.fr

L. Magalhães and J. Goldstein (eds.), *Toys and Communication*,
https://doi.org/10.1057/978-1-137-59136-4_7

It could be thought that toys belong to the world of childhood. But, in reality, they invite us to look more closely in order to 'see more,' just as Blake suggested in his *Auguries of Innocence*. That invitation, 'to see a World in a grain of sand' becomes even more urgent when the toy emerges as an image in the history of art. Even though we are constantly looking, in a society full of images, it is not easy to see what is really there. Indeed, the act of looking is more demanding than the act of seeing. Delving into the exhibition rooms of an art museum, we look at paintings, sculptures, and antique prints. But do we really see the entirety of these works and the details therein? If looking is, in essence, 'an act of knowledge' (Pieper 1999: 191) then the poet Konrad Weiss is right in saying that: 'contemplation does not rest until it has found the object that dazzles it' (Weiss 1990: 41). Could the toy, then, be one of those objects that art historians have not known how to see?

To remedy this old blindness, we must contemplate the artistic images of each epoch and try to see that which the icon 'reveals but does not exhibit' (Lafuente Ferrari 1985: 148). The toy in art is the trace of a message added to that which the painting itself, at first sight, presents to us. As Roland Barthes intuited: 'What is hidden is, for us Westerners, "truer" than what is visible' (Barthes 1980: 156).

This 'hidden' element which throbs in the minor detail of a painting is that which the celebrated art historian Kenneth Clark set out to explore in a book entitled *One Hundred Details from the National Gallery* (1938). He compared, for example, the young cupid painted by Velázquez in *Venus at her Mirror* with the little angels in *The School of Love* by Correggio, revealing fascinating analogies. Clark's essay became a publishing phenomenon that led him to publish a second volume, *More details from Pictures in the National Gallery* (1941). Although the proposal turned out to be less attractive to art scholars than it did to a public spellbound by the photographic extensions of the book, his intuition did not fall on deaf ears. Fifty years after that first analysis undertaken by the then director of the National Gallery in London, Daniel Arasse published *Le Détail: pour une histoire rapprochée de la peinture* (1992). In this richly illustrated essay, the French historian invites us to consider how the details of a painting reveal the personality of the artist. In the smallest detail, there is usually a hidden message. Sometimes the author introduces it in a deliberate way; while in others he leaves it to the social imagination of its epoch. It was for this reason that José Ortega y Gasset referred to the moment of creation as 'the well-known.' These details may well seem obvious to the artist and their contemporaries; however, since they become out-dated, later ages find themselves in front of a 'half-alien culture' (Burke 1998: 4) and so have to do some deciphering.

It is at this point that our analysis of the toy in Spanish art from the fifteenth to the nineteenth centuries begins to make sense. Previous investigations having explored the case of French, Italian, German, and Flemish painting (Manson 2001; Vaz-Romero Trueba 2011), Spanish art remains unexplored in respect of the representation of the toy. First of all, it cannot be considered an accidental phenomenon that toys emerged in Spanish art: just as this was not the case in the rest of the late-medieval and modern European context. Introduced into a complex composition, the toy-as-image takes on a meaning that is distinct from that of the toy-as-artefact. Its mere presence becomes language, and the attempt to translate that language is itself the 'reward' that Arasse speaks of. For, analyzing the toy-as-image, we understand that: 'a work of art is built up of a hundred different sensations, analogies, memories, thoughts—some obvious, many recondite, a few analysable, most beyond analysis' (Clark 2008: 8). Indeed, in his study of the representation of life and everyday objects in Flemish Renaissance painting, Tzvetan Todorov warns:

> The difference with regard to the past is not that these objects or actions were not represented before, but that they were previously considered not enough by itself as the organizing principle of the painting. However, after this time [sixteenth century] they can get to be. The accessory attribute has become increasingly the essential status, and what was subordinate has become autonomous in the history of painting (Todorov 1993: 10).

From this observation, it follows that the representation of the toy is 'praise' for the object on the part of the artist, and a positive recognition of the reality of the child. Despite some unremarkable exceptions, this phenomenon spread throughout Europe from the fourteenth century. It was in the following century, however, that this propagation accelerated, and in the case of Spain, it really began to stand out in the early sixteenth century. So it was that the toy would gradually come to be included in different media and artistic genres, thus giving way to a notable semiotic branch. The images of the early centuries, few in number, lose their religious code to take on more secular and narrative iconographic formulas, over which the explosion of the toy in the art of the twentieth century would be consolidated.

An Allegorical Language: The Toy as a Moral Emblem

In the late Gothic period, this 'praise' of the toy does not arise in the industrious workshops of the painters of Castile and Flanders. It comes about, however, slowly and discreetly, in the margins of illuminated manuscripts, whose satirical character has been studied at length (Randall 1966; Camille 1992; Wirth 2008). These miniatures would incorporate two models proceeding from European scholastic pedagogy: the *puer senex* or 'old boy' and the simple, mundane *infans*: the first, seen as being Christian and holy in his play, being opposed to the second, weak in body and spirit, and prone to crazy unreason (Vaz-Romero 2016). This ambivalent consideration is that which appeared in some manuscripts and incunables. When toys occupy the main illustration on a page, praise of children's play is what prevails. But when the author places them in the ornamental friezes of the page, known as 'burlesque margins,' the image serves to criticize an inverted and infirm world, where the figure of the child is confused with that of the crazy, the heretic, the wild, and the monstrous. All these are united by the facts of being creatures in the 'margin' of society and of divine redemption.

In a Catalan book of hours from the fifteenth century, the illustrator has represented the toys of Jesus, evoking a childhood which is presented as that of an ordinary child (Fig. 7.1b).[1] The Christ Child plays in the carpenter's workshop of Nazareth, under the watchful eye of his father. According to this image, it would not be unlikely that Joseph had made the toys with his own hands, since the ball and the tops are of finely turned wood. The boy also has a living toy: a warbler tied by a silver thread. On the table, a pear symbolizes the sweetness and softness of Mary's feelings. On the floor, next to the toys, some humble ants, austere and virtuous, and working secretly, represent the model of a good Christian. There are seven of them: the sacred number, signifying evangelical perfection, and the imitation of the humility of the Holy Family. The miniature portrays the pedagogy of Saint Jerome and Saint Thomas Aquinas; thus legitimizing the mundane play of the Incarnate Word, whose time of messianic sacrifice has not yet arrived. The toys are well integrated within the scene of domestic bliss, which is in itself a blessing, as foreshadowed in Psalm 128.

Nevertheless, on the front of the Navarre altar, dating from the second quarter of the fourteenth century, and principally depicting scenes from the childhood of Jesus, we discover what could be a toy with a very different

◀ **Fig. 7.1** **a** Child with flower and windmill. Drawing by Oriol Vaz-Romero Trueba (ink on paper, 21 × 25.3 cm) based on: *The Childhood of Jesus*, front panel with a frieze of the Seasons. Parochial Church of Arteta, Navarra (c. 1325–1340). Tempera, gold leaf/wood, 90.8 × 171.2 × 5.8 cm. **b** *The Holy Family: Christ Child playing with wooden balls and tops*. Drawing by Oriol Vaz-Romero Trueba/ Esther Alsina Galofré (i/p, 29.7 × 21 cm) based on: A detail of a Catalan Book of Hours (15th century). **c** Wild child playing with a windmill. Drawing by Oriol Vaz-Romero Trueba/Esther Alsina Galofré (i/p, 28.5 × 17.8 cm) based on: *Comentaria Jacobi de Marquilles Super vsaticis barchino* (1505). Barcelona, Joan Luschner [Impr.], f° 1r. © OV-RT.>>

meaning: a boy holding two pinwheels, like flower stems, representing the climatic changes of April (Fig. 7.1a).[2] It is precisely this toy, symbolizing the whim of the winds, which, within the margins of illuminated manuscripts, serves to define the defects of childhood. In the first folio of an incunable of commercial laws printed in Barcelona in 1505, the honorable legislators of the city are represented in the main image, while the variegated page margins show an upside-down world where debauchery and animalism prevails (Fig. 7.1c).[3] Among other less virtuous characters, the woodcut of the printer Joan Luschner presents a child riding on a branch. They brandish a pinwheel with four blades, and are in a fighting stance, pointing forward. It is one of the most popular toys and the one which has been most depicted in the medieval iconography of infantile play. It is a symbol of *infirmitas*, an attribute of pre-man and therefore a negative moral emblem: it turns and turns, rocked by the wind, as the child is carried away by his carefree and whimsical game.

The hobby-horse is another popular toy from the time of Horace: 'equitare in arundine longa' (*Satires* II, 3, 248), and it also expresses an allegorical ambivalence. In a German theological manuscript from the early-sixteenth century the baby Jesus is depicted riding a hobby-horse, brandishing a rod and challenging the snake.[4] From an iconographic point of view, this toy, often made by the children themselves, or a product of local craft, becomes a weapon in the service of light to combat the darkness of evil, represented by the serpent. However, in an illustration of the *Emblemas Morales* (*Moral Emblems*) by Sebastián de Covarrubias (1610), the artefact symbolizes the folly of children and the elderly (Fig. 7.2a).[5] Just as the emblem has it: 'Bis Pveri Senes,' which is to say, 'Old men are twice children.' Covarrubias' proverb leaves no room for doubt: the elderly, who do not possess the wisdom of their own age, are doubly

BIS PVERISENES

◀ **Fig. 7.2** **a** Bis Pveri Senes. Drawing by Oriol Vaz-Romero Trueba (i/p, 21.3 × 28.5 cm), based on: Sebastián de Covarrubias Orozco (Madrid, 1610), *Moral Emblems*, C. I: 91. **b** Putti riding hobby-horses. Drawing by Oriol Vaz-Romero Trueba (i/p, 21 × 29.7 cm), based on: Juan de Bruselas (1512–1516). Carved wood, 38.7 × 52.5 cm. Misericord or mercy seat, Choir of Zamora Cathedral. © OV-RT

children (Langmuir 2006: 136). The Spanish lexicographer returns to a *topos* dating back to Latin literature, in particular the aforementioned satire of Horace and the *Stromata* of Clement of Alexandria (VI: 2), who in turn takes up texts by Sophocles, Antipho, Plato, and Theopompus (Roberts and Donaldson 1869: 314; Mateo Gómez 1979: 189). This theme was also one of the most recurrent throughout the late-medieval period and the Spanish Golden Age (Bouzy 1993: 132–133).

It is striking that this proverb evolved from ancient times to reach the *marginalia* of medieval manuscripts, always having as a recurring element the hobby-horse. Its popularity even led the sixteenth-century master carvers to depict them on choir stalls across Europe, especially in the north. In France, they appear on the mercy seats of Saint Anne of Gassicourt, Notre-Dame of Amiens and Saint-Lucien-les-Beauvais. In the Spanish Netherlands, we find examples of children riding hobby-horses on the mercy seats of Hoogstraten (Maeterlinck 1910: 202, 204), on one of Dordrecht and even on the floor of the choir of the Church of Saint Sulpice in Diest (Bethmont-Gallerand 2002). One influence taken from Flemish artisans is the child riding a toy horse, possibly made by the sculptor Juan Guas in the late-fifteenth century, on one of the columns of the Gothic cloister of San Juan de los Reyes, in Toledo.

Furthermore, the mercy seat of the Cathedral of Zamora, carved by Juan de Brussels between 1512 and 1516 (Teijeira Pablos 1996), depicts three children mounted on hobby-horses, one of whom also wields a pinwheel with two blades (Fig. 7.2b). On the mercy seat of a Salmantinian cathedral in Ciudad Rodrigo, the carver engraves an old man riding his toy, as the emblem of Covarrubias.

This playful old man follows a pattern identical to that described by Sebastian Brant in his famous *Das Narrenschiff* or 'The Ship of Fools' (c. 1494), which satirizes the passion of the elderly to imitate children. It would not be strange that Cervantes also let himself be seduced by the allegories of Brant and Covarrubias, having his old-child Don Quixote mount a wooden horse called 'Clavileño the Swift' (II, 40–41), while blindfolded as an allusion to his lack of judgement. The corruption of the secular clergy, the morality of the private life of the masses, and the religious crisis were largely caused by the spiritual chaos of the Western Schism, justifying the emergence of secular iconography on Spanish choir stalls (Matthew Gómez 1979: 30–31). The intention was simply to highlight the damage caused by vice, with the moral *ad terrorem* being the corrective by which to avoid it.

It is not surprising, then, that the pinwheel and hobby-horse serve to illustrate the stages of life, as in the theological treatise of Antonio de Honcala (1546). In the woodcut which presides in his *Pentaplon Christianae Pietatis*, Juan de Brocar sketched at the bottom of the road of life a young baby next to a little boy playing with a pinwheel and a bird tied like a 'living comet.' Just above, a Latin phrase helps the detail to be interpreted: 'Heus iacet in guild tetre infans caliginis. Ignarus rerum nescius ipse sui,' which is to say: 'Unfortunately, the baby is wrapped in Darkness, because he ignores the World itself.' The two paths on which the child with its toys is found lead us to understand that it is a being whose steps can lead to vice or virtue. It is a recurring graphic approach that is repeated even in the *Tabula Cebetis* (The Table of Cebes) of the sixteenth and seventeenth centuries (Fig. 7.3a). In Spain, these tables lent their name to many allegorical treatises, such as that of Ambrosio de Morales, taken up again in 1672 by the Brussels editor F. Joannes Foppens, who adorns it with a fold-out page made by the engraver Matthäus Merian (Pedraza 1983; López Poza 2001). In this, as in the first Spanish versions of Juan Martinez Population (1532), Juan de Jarava (1549), and Pedro Simon Abril (1586), the composition is dominated by the mountain of wisdom, at the foot of which is always found an infant, helpless and absorbed in its play. Even in the chalcographies in the editions of the eighteenth century, such as *Paráfrasis árabe de la Tabla de Cebes*, translated in 1793 by Pablo Lozano and Casela, the Madrid engraver J.L. Enguídanos again takes up a detail repeated ad nauseam: girls always play with dolls, and boys with balls, spinning tops, wooden horses, and pinwheels (Fig. 7.3a).[6] Various writers and engravers place children away from the holy mountain, outside the walls or before a fork in the road. It is a graphic way of saying that they are creatures who still ignore the roads to the City of God. This tradition would be well rooted in the Spanish cultural imaginary, as is shown by the many decorative recipients made in the potteries of Talavera de la Reina (Toledo) and certain allegorical works such as the portrait of the jester Calabacillas, painted by Diego Velázquez,[7]

the drawing of José de Ribera entitled *Boy with a pinwheel and an old man pulling a cart with a corpse,*[8] or the sketch of the *Weeping Child*, upon which Goya bestows a pinwheel.[9]

◀ **Fig. 7.3** **a** Children playing with a rattle, a doll, a hobby-horse, and a windmill. Drawing by Oriol Vaz-Romero Trueba (i./p., 31 × 16.7 cm), based on: Matthäus Merian, engraving accompanying a Spanish Tablet of Cebes included in the *Theatro Moral de la Vida humana* by Ambrosio de Morales (Brussels, F. Foppens, 1672). **b** Isabel of Spain with a doll. Drawing by Oriol Vaz-Romero Trueba/Esther Alsina Galofré (i./p., 23.8 × 16.4 cm), based on: A detail of *The Stoneleigh Triptych*, 1506. Oil/panel, 32.5 × 15 cm. **c** Future Carlos I playing with a toy propeller. Drawing by O. V.-R./E. A. G. based on: A detail of *The Preaching of St John (Lc. 3, 1-6)*, 1515–1520. Silk, silver, gold, wool, 353 × 409 cm. © OV-RT

These emblems are the expression of an intellectual discourse. As we have seen, they require a title, a legend and even a prepared text in order for their meaning to be deciphered. In short, the toy 'speaks' a hermetic language to adults. It is the codification of a foolish age, especially among children, but also for the insane elderly. On other occasions, the artist decides to turn play into a 'future promise,' in an early sign of holiness. The mill and the wooden horse for boys, and the dolls for girls, as shown in a lunar emblem of the Pisanello School (1440),[10] also draw sexual differentiation in the iconographic attribution of the toy, unveiling a new semantic territory of the object.

A Gendered Language: Toys Conquer the Canvas

Throughout the sixteenth century, a more accurate gaze falls on the child (Manson 2001: 59–82, 2007). That is when the toy becomes an informant of the subject of the portrait and the expectations that adults place on it. The toy imposes a gender identity (Manson 2004), anticipating the functions associated with adulthood. Toy soldiers point towards a male education, orientated towards military capacities and government. Meanwhile, china dolls and tea sets announce roles which are distinctly female: motherhood, household management, fashion, and adult pastimes in society. With this formalized gendered, in the princely portraits, later emulated by the bourgeoisie, the iconography of the toy lightens its moral burden to embrace a social meaning. It symbolizes the mission of all those children portrayed not for their own sakes, but for being the main guarantee of the continuity of a heritage that surpasses them. On the other hand, and without us having noticed it, this mechanism has allowed the toy to begin to conquer a new artistic medium: the canvas. So it is that the toy loses the marginality of the medieval emblematic to become the main composition of the work. This also enables it to be painted on a larger scale and in more detail.

The triple portrait known as the *Stoneleigh Triptych*, produced in 1506 by an anonymous painter associated with the Castilian court, depicts the three children of Archduke Felipe of Habsburg and Queen Juana I of Castile: the future emperor Carlos, Eleanora of Portugal, and Isabel of Austria. Little Isabel carries in her hands a richly dressed doll, which indicates the girl's future and the main role that she will have to assume (Fig. 7.3b).[11] The infant will not be regent, but matriarch. She will marry King Christian II of Denmark. He will renounce political power in Spain, but will ensure dynastic continuity by providing for himself new heirs in distant lands (Marsillach 1944). In contrast, the tapestry woven for Queen Juana I of Castile shows Carlos, Isabel's brother, playing with a toy propeller whose symbolism, linked to the sphere, is more complex than that of the pinwheel (Fig. 7.3c).[12] The tapestry, woven in the looms of Pierre van Aelst and following the cartoons of Jan Van Roome, shows a special detail in the face and in the play of the child, situated amidst the preaching of Saint John (Herrero Carretero 2004: 58, 103). As in the French paintings of the same epoch, which represent the baby Jesus, this toy propeller is the image of power over the world. It is significant that the artist has placed this power in the hands of the child, since the date of manufacture of the tapestry, between 1515 and 1520, is close to the proclamation of Charles as King of Spain and the Holy Roman Empire.

In Isabel's case, the hieratism of the doll evokes docility as a mission, while the blades of the toy propeller of Carlos claim the ability to move and control. This pattern is repeated many times, as is shown by the paintings of royal Spanish infants from the sixteenth and seventeenth centuries. The Infant Don Diego, son of Philip II, portrayed at the age of two years by Alonso Sánchez Coello,[13] holds in his left hand a wooden spear and a luxurious hobby-horse, designed to be used by a young prince. In turn, a piece of canvas painted by an anonymous Spaniard in 1600 shows a young heiress with her doll. The piece was trimmed and reused in later centuries as a fireplace screen in Castle Drogo in County Devon, following the eccentric tradition of 'silent companions' or dummy board figures (Fendelman 1981; Edwards 2002).

The conquest, first of the altarpiece and later of the canvas, brings more realism and a more varied knowledge of existing recreational artefacts, especially those of trade and commerce. Thanks to portraits such as Santiago Morán's, of the Infant Margarita Francisca (1610),[14] daughter of

Philip II, and Diego Velázquez's, of a young Baltasar Carlos (1660),[15] son of Philip IV, we can trace the exact appearance of the rattles of royalty. Gold, silver and coral were the most commonly used materials. But rattles did not last long in the hands of infants. In the portrait of Baltasar Carlos, Velázquez captures the moment in which the dwarf, the infant's playmate, is quick to take the rattle and replace it with a baton: nothing must be permitted to remind the prince of his childhood. The time has come for him to prepare to be inheritor of all the estates of the Spanish monarchy. Velázquez and his workshop represent him at the age of six years old sporting garments identical to those of adults, with tabard, lopsided cap, or suits with brocades of gold, standing within a hunting context. On two occasions[16] he will even be painted with a hunting musket and a little sword made to measure. As J. Portús says, 'hunting is the image of war' and would therefore constitute a context that very appropriately referred to the military responsibilities that awaited the future king (2006: 96).

This language, connotative of gender and the education of children, still exists in the early-nineteenth century. So it is that the court painter Antonio Carnicero conceives the portrait of the Infant Maria Luisa Carlota of Bourbon (c. 1804),[17] who carries a delicate toy carriage. The doll is lacking, but the artist manages to express the caring motherhood that the court expected for this girl. However, in the portrait of one of the children of Charles IV, dated between 1798 and 1802, Carnicero is quick to give the child a fitting toy: a tambourine and a small battle drum. Another example is the wooden chariot, richly crafted and painted with green and gold highlights, as captured by Francisco de Goya (Fig. 7.4a).[18] This is also a princely toy; however, it was hardly ever represented in European painting, and consequently allows the painter of Fuendetodos to introduce a mark of distinction in the figure of little Pedro. In the family portrait of the Duke of Osuna, Goya also introduces an improvised toy. Francisco de Borja holds between his legs a sheathed sword like a palfrey, as if he is emulating the military exploits of his father, Pedro Tellez Giron, the Duke of Alcántara.

In his role as court painter, Goya continues the codifying tradition of Velázquez and Carnicero (Beruete and Moret 1916: 36–38). In his portrait of Pepito Costa Bonells (1813), son of King Ferdinand VII's physician and grandson of the family physician of the Dukes of Alba, we notice a drum, a rifle, and a wooden bayonet.[19] But what most stands out is the 'big' black horse on wheels, whose composition is reminiscent of the so-called dismounted equestrian portraits, popular among military figures

◀ **Fig. 7.4** The heirs of Osuna with their toys. Drawing by Oriol Vaz-Romero Trueba/Esther Alsina Galofré (i./p., 21 × 29.7 cm), based on: Francisco de Goya (1787–1788), detail of *The Dukes of Osuna and their children*. Oil/canvas, 225 × 174 cm. Drawing by Oriol Vaz-Romero Trueba/Esther Alsina Galofré (i./p., 23 × 20 cm), based on: Francisco de Goya (1778–1779), *Boys playing soldiers*. O./c., 146 × 94 cm. © OV-RT

and royalty during the seventeenth and eighteenth centuries. All these toys, along with the soldier's jacket and the Napoleonic hairstyle, seem to allude to the Spanish War of Independence (1808–1814), but also acquire an augural character as Pepito will go on to join the militia of the Madrid infantry.

Numerous aristocratic portraits demonstrate the role of toys in gender differentiation. The romantic artist put into the hands of future ladies of the kingdom all manner of hoops, balls, stuffed animals, dolls, and dishes, while men are provided with an arsenal of drums, wheels, or rocking horses, miniature soldiers, and armoured vehicles or transport. This is demonstrated by the portraits of children by Agustín Esteve y Márquez, Rafael Tegeo Díaz, José Elbo Peñuelas, Antonio María Esquivel, José Roldán Martínez, José María Romero López, Angel María Cortellini, Augusto Manuel de Quesada, Victor Manzano Mejorada, and Raimundo de Madrazo, to name but a few of the painters who most applied the aristocratic model to the children of the burgeoning middle class. These portraits do not involve an awareness of the uniqueness of the child, of his personality through play. In their bewitched face, in the stiff composure of the body, in the solemnity of their clothing, and in the kind of toys that appear, a well-established political and social message is expressed with the intention of affirming the reputation and continuity of a well-ingrained lineage. Even when the artist portrays his own children or children of their family environment, the old coding system is utilized. Only on rare occasions are expressive elements permitted to be seen.

This is the case with Francesc Torras Armengol, in the portrait of his daughter.[20] The face of the girl and the maternal way in which she embraces her doll betray an attempt to capture the personality of the subject. In a way, there is an underlying pre-narrative element, as if hugging the doll tells a story that only the artist, as a parent, knows. The same can be said of a painting by José María Romero López, in which he portrays the brothers of the Santaló family.[21] Like any good romantic, the Andalusian painter constructs an atmosphere of sorcery. Influenced by the

Biedermeier and Victorian styles, Romero López places the toys in a lavish antechamber, decorated with mirrors and small religious scenes, and leading out onto a blue and misty garden. Others, like Manzano Mejorada,[22] would position toys next to the fireplace of the family living room. José Borrell del Caso would choose, in turn, the forest clearing of Puigcerdà Valley.[23] Slowly, the emotional relationship that connects the painter and the child with their toys would prevail over the symbolic language of identity of previous centuries. From that emotion, perceived by the artist, would be born a new way to capture children and their playful universe.

A Narrative Language: The Toy and the Dramatization of Space

The 'narrativization' of the compositional elements is conceived in the painting of the Spanish Golden Age, and slowly starts to be introduced into later childhood portraits until it reaches the 'costumbrist' scenes generated in the athanor of late Spanish romanticism. One of the most illustrious predecessors is Bartolomé Esteban Murillo. The master of the Sevillian School would be one of the first to bring to the canvas the street scenes of child 'rogues.' Unlike the princes and royal children, these children without relations or legacy are painted as individuals of flesh and blood and not simply as heirs to a lineage. These dirty, hungry, playful, and feisty creatures conquer the painting as a result of an emerging taste for picturesque and profane subjects. This is self-evident in the canvas painted by Murillo between 1670 and 1675 in which there appears an elderly woman in her humble abode, delousing a child, perhaps her grandson.[24] The youngster is busy playing with a dog as he clutches a piece of bread in his left hand. Nevertheless, the rogue has reserved for after the 'toilette' a beautiful wooden horse, perhaps stolen from a Sevillian child of the nobility.

From the landscape tradition popularized in Rome by Annibale Carracci, Claude Lorrain, and Nicolas Poussin and his followers, the taste for the picturesque and sublime strongly takes root in the canvases of the eighteenth century, offering the child and his play new iconographic scenes. A landscape with ruins painted around 1773 by José Carlos de Borbón, a black artist who came to Spain from Naples in 1759 in the entourage of King Carlos III, is one of the first instances in which the toy appears integrated in a narrative scene.[25] What certain conservators have confused with a toy guitar (Espinós 1986: 53) is actually the classic pinwheel with two blades present in medieval images. One of the boys plays with a woman, perhaps his mother or his nurse, close to a fountain and some classically inspired ruins, wrapped in twilight. From the same time is a representation of *Madrid Fair in Cebada Square*, where Manuel de la Cruz Vázquez does not fail to paint, in the midst of the crowd, under an expansive sky pierced by the roofs of houses, a boy with the same toy that we find in the landscape of José Carlos de Borbón.[26] But it would be José del Castillo, painter and engraver of the Royal Tapestry Factory of Santa Barbara in Madrid, who first cast his gaze on the popular games of Spain. We find several scenes with children of the working classes or the bourgeoisie playing with balls, spinning tops, and bowls in an idyllic, outdoor setting, far removed from anything remotely urban. These pieces, commissioned on April 28, 1780 at the Royal Factory, are part of a series of five scenes intended to decorate the vanity table of the Princess of Asturias, the future Queen María Luisa, in the Royal Palace of El Pardo.[27]

Francisco de Goya, also under contract to the Royal Tapestry Factory from 1780, would take inspiration from the work of his rival, José del Castillo, and would produce several series of cartoons depicting children's games for tapestries to decorate the rooms of the Princes of Asturias in El Pardo and El Escorial, as has been reconstructed by the watercolor artist F.J. Hernández Alonso (Rosenthal 1982; Herrero et al. 1996). More than a dozen 'country themes' with street children playing in the suburbs and mountains of Madrid recapture elements of the late-medieval emblematic and of the chalcographies produced by Claudine Bouzonnet-Stella for the edition of *Les Jeux et Plaisirs de l'enfance*, compiled by his uncle Jacques Stella in 1657. Like many of his European contemporaries, Goya must have been familiar with the illustrations of some of the editions of Stella, if we take into account the fact that his inventory of goods, compiled on October 28, 1812 on the writing desk of Antonio Lopez de Salazar, consists of books, in Spanish and foreign languages, worth 1500 Spanish reales (Mestre 1990: 25–34).

One of Goya's most striking compositions features children playing soldiers with their bayonets and toy drums (Fig. 7.4b).[28] Alongside them, another little boy is having fun with a mysterious miniature bell tower, leaning against a wall. What is the significance of this toy, of which we do not have any historical record as a plaything? Maybe it is a symbol through which the artist tries to send a message to the young Princes of Asturias, as if saying to them: 'The bell touches the deceased, announcing the end of a regime.

Look, children of kings! Behold the games of the common children. Know your people if you want to keep your crown.' Indeed, the collection of Goya cartoons shows groups of rascals playing at bullfighting[29] and fighting amongst themselves over chestnuts.[30] These scamps, somewhat violent, scare the little girl in the foreground, who tries to escape with her little wooden cart. Other boys, located in a forest clearing, are enjoying inflating pig bladders,[31] playing with a rocking-horse with stumps,[32] or dragging a cart to the sound of piccolos and little drums.[33] Perhaps they are imitating the great triumphal chariot processions of Emperor Maximilian I, made famous by the *Triumphzug* by Dürer and Albrecht Altdorfer, and reproduced to excess by their followers.[34] Not in vain is a similar scene chosen in 1846 by the portrait artist and historical painter Luis de Madrazo y Kuntz, placing his four nephews like participants of a parade, and with a richly adorned doll.[35]

From the eighteenth century, under the influence of philosophers of education such as Comenius, Locke, and Rousseau (Manson 2001: 125–154, 197–230), the child occupies a new space in the ideas of European culture. Children begin to be represented for their own sake and not for their social position. This is the result of the romantic myth of childhood and, by extension, play (Vaz-Romero 2011: 1623–1805). It is true that this imaginary took root later in Spanish culture compared to a German, English, or French context: a 'delay' that is due to a longer survival of the courtly style in the pictorial commissions of wealthy families of the nineteenth century. But the taste of the civil bourgeoisie would prevail and, with it, their social models. The paintings praised the rise of material goods and the fashions of domestic life, so the artists can allow themselves to truly see the child playing. Both the artist and the patron of the painting find a new fascination in this little human being and its toys. Whether constructed by the artists themselves, or originating in commerce, recreational artefacts do not cease to increase in number, due to the birth of the production line and the consequent increase in productivity.

Sometimes this look is expressed ironically, as it is by Francisco Díaz Carreño in *Probable Position of the Globe before the Great Flood* (Fig. 7.5a).[36] A girl is about to spill the water from a fishbowl, along with the fish inside it, onto the bedroom floor. It is her own personal 'small Deluge': more exciting than the beautiful mechanical toys such as the train and the tricycle, and by it Díaz Carreño suggests a romantic critique of the industry. The same happens in the painting by Ramón Martí Alsina representing his son Ricard, around 1870, at the age of 12.[37] Bored with his

toys and all his presents, the boy sinks into an armchair in the library, dreaming of a different life from the one that he is forced to lead. It is the same critique made by E.T.A. Hoffmann in his famous *Das Fremde Kind* (1817): the caricature of the modern child who can no longer play in the street like the naughty boys of José del Castillo, Goya or Ramón Bayeu.[38]

◀ **Fig. 7.5** **a** Some toys of an urban girl in the industrial era. Drawing by Oriol Vaz-Romero Trueba/Esther Alsina Galofré (i./p., 26.6 × 21.3 cm), based on: Francisco Díaz Carreño (1890), *Probable position of the Globe before the Great Flood.* O./c., 145 × 100 cm. **b** Drawing by Oriol Vaz-Romero Trueba/Esther Alsina Galofré (i./p., 23 × 16.2 cm), based on: Joan Ferrer Miró (1889), *Bringing toys.* O./c., 73 × 52 cm. **c** A girl asks Saint Peter to enter Heaven with her doll. Drawing by Oriol Vaz-Romero Trueba/Esther Alsina Galofré (i./p., 21 × 29.7 cm), based on: Josep Maria Tamburini (1898–1904). *Celestial Story.* O./c., unknown size and location. © OV-RT

The city grows in size and in danger. The domestic space and the toys bought in the brand-new city stores are the only possible entertainment for well-off children in the late-nineteenth century. This urban setting was conducive to a variety of narrative scenes featuring toys, many of which would require specific studies exceeding the brevity we have set ourselves in these pages. As examples, we can point to the painting by José Borrel del Caso called *Neighbours on a roof in Barcelona* (1891), *The first pair of trousers* by Lamberto Alonso Torres (1897),[39] and *A Winter Evening* by José Mongrell Torrent (1899).[40] In *The Breakfast of the Painter's Nephews with toy soldiers*, from the same epoch, the toys painted by Darío de Regoyos, unlike the impending *flood* of Carreño Díaz, portray a game that has finished in the comfortable family living room.

Most of these paintings belong to private collections and have barely been studied. A set of themes related to Christmas and the gifts of the Three Kings also exists, which adds a poetic character to the aforementioned urban scenes. As we have shown, in the late-nineteenth century the Christmas and Epiphany festivals turn into celebrations for children not only from a religious point of view but also from a commercial one (Manson 2005). Joan Ferrer Miró portrays himself accompanied by his son in the streets of Barcelona, carrying the Epiphany gifts, among which we distinguish a magnificent painted cardboard horse with wheels and a mane of hair (Fig. 7.5b).[41] In a canvas exhibited at the Paris Exposition Universelle, 1889, the Catalan artist focuses again on the commercial side of the Epiphany, portraying the window of a toy store in Barcelona. He tries to capture the many parents and children who crowd the doors of the toy shop, some coming out with their treasures and others, impatient, jostling to get in, since there are so few hours in which to prepare their gifts at home, by the fireplace or next to the Christmas tree. Others, like Joaquín Sorolla, represent the very moment that children, overseen by their mother, place their shoes next to the window, so that the Three Wise Men know to give each one the corresponding mysterious gift.[42]

Numerous paintings and prints which appear in the illustrated newspapers of the time feature images of children dreaming during the night before Kings' Day of the toys they may receive at dawn. For these scenes, painters like Ramon Tusquets,[43] J. Diéguez,[44] and Josep Cusachs[45] turn to the magical language of fairy tales and the allegories bequeathed by the Christian tradition (Vaz-Romero 2014). One triptych stands out. Now lost, it was painted by the symbolist Josep Maria Tamburini, and features the supernatural encounter between a little girl and Saint Peter (Fig. 7.5c).[46] She hides behind her beloved doll, soliciting for it a place in paradise. All these images are dominated by movement, body gestures, and the need for a literary text. Therefore, the three languages explored, the allegorical, the identity, and the narrative, do not occur in time as if they were watertight compartments; instead, they continue to exist across those periods in which new semantic formulas for the image of the toy arise. They intermingle, encouraging one or another language, but always remaining in the cultural imaginary of Spain.

In short, we have seen the toy, that tiny symbolic object, 'speaking' in an allegorical language about how man comes to this world, passing first through his impulse for recreational passions before attaining true knowledge on the mountain of life. The toy's iconic presence, born in the margins of Gothic manuscripts, becomes more and more central and significant in art, to the point of becoming part of the subject of paintings. At the same time the child is individualized; he becomes a person and, finally, the actor of his own life. That is how the toy in Spanish art at the doors of the twentieth century, however little we decode the messages it presents in each epoch, has told us how society, over the course of six centuries, has changed its way of looking at the child and its inner world.

Notes

1. London: British Library [Add MS 18193, f° 48v].
2. Barcelona: Museu Nacional d'Art de Catalunya [004368-000].
3. Barcelona: Biblioteca de Catalunya [347 (467.11)Mar].
4. Stuttgart: Württembergische Landesbibliothek [*Cod. Theol.* 4° 136, f° 19].
5. Madrid: Biblioteca Nacional de España [ER/1334].
6. Princeton: Princeton University Library [(Ex) B561.E6 E5 1670 copy 2].
7. Cleveland: Cleveland Museum of Art [Accession Number = 141 549].

8. Madrid: Museo Nacional del Prado (MNP) [D08551].
9. MNP [D04060].
10. Modena: Biblioteca Estense Universitaria: *Liber physionomie* [MS Lat. 697 = α.W.8.20, f° 4v].
11. Mallorca, Fundación Yannick & Ben Jakober [N. JS1].
12. Madrid: Palacio Real [10005831].
13. Alonso Sánchez Coello (c. 1577). *Retrato del infante Don Diego*. Oil/canvas, 108 × 88.2 cm. Vienna: Liechtenstein Museum. A studio version is in the Alte Pinakothek, Munich [4199].
14. MNP [P1282].
15. Boston: Museum of Fine Arts [01.104].
16. MNP [P01189; P01233].
17. MNP [P4718].
18. MNP [P739].
19. New York: Metropolitan Museum of Art [61.259].
20. Francesc Torras Armengol (c. 1865). *The artist's daughter*. O./c., Ø 44 cm. Museu de Terrassa.
21. Madrid: MNP. Inv. P04608.
22. Madrid: MNP. Inv. P03976.
23. José Borrell del Caso (1896). *Portrait of Mr. Moner's son in Puigcerdà Valley*. O./c., 123 × 83 cm. Private Collection.
24. Munich: Alte Pinakothek [C447 M216].
25. MNP [P05287].
26. MNP [P00693].
27. MNP [P03311; P03313].
28. Madrid: Museo Nacional del Prado [P00783].
29. Madrid: Fundación Lázaro Galdiano [2593].
30. Madrid: Colección Santamarca.
31. MNP [P00776].
32. Philadelphia: Ph. Museum of Art [1975-150-1].
33. Toledo (Ohio): Toledo Museum of Art [1,954.14].
34. Vienna: Albertina [25246]. Cf. Vienna: Österreichische Nationalbibliothek, Cod. min. 77, f° 2, 5, 10.
35. Madrid: Colección Madrazo [Madrazo-77].
36. MNP [P05574].
37. Barcelona: Museu de Montserrat [N. R. 200.455].
38. MNP [P02599; P03071].
39. MNP [P06735].
40. *Velada de invierno*. José Mongrell Torrent, 1898–1901. O./c., 57 × 96.5 cm. Private Collection.
41. Barcelona: Private Collection.
42. Madrid: Museo Sorolla [00486].

43. Ramon Tusquets (c. 1881). *Allegory of Time.* Watercolor, 73 × 55 cm. Private Collection.
44. *Año Infantil* (6-I-1896). Barcelona: Tipolit. Henrich & Ci[a].
45. *Album Salón*, N° 8 (9-I-1898), p. 11.
46. The original painting was reproduced in: La Ilustración Artística, N° 1155 (15-II-1904), p. 128.

References

Arasse, D. (1992). *Le Détail: pour une histoire rapprochée de la peinture.* Paris: Flammarion.

Barthes, R. (1980). *La Chambre claire: Note sur la photographie.* Paris: Éditions de l'Étoile, Gallimard, Seuil.

Bethmont-Gallerand, S. (2002). La joute à cheval-bâton, un jeu et une image de l'enfance à la fin du Moyen Age. *The Profane Arts of the Middle Ages, 9*(1–2), 183–196.

Beruete y Moret, A. (1916). *Goya. Pintor de retratos.* Madrid: Blass.

Bouzy, C. (1993). L'emblème ou le proverbe par l'image au Siècle d'Or. *Paremia, 2*, 125–134.

Burke, P. (1998). *The European renaissance: Centres and peripheries.* Oxford: Blackwell.

Camille, M. (1992). *Image on the edge: The margins of medieval art.* Cambridge: Harvard University Press.

Clark, K. (1941). *More details from pictures in the national gallery.* London: Trustees by Harrison & Sons.

Clark, K. (2008). *One hundred details from pictures in the national gallery.* London: National Gallery, Yale University Press.

Edwards, C. (2002). Dummy board figures as images of amusement and deception in interiors, 1660–1800. *Studies in Decorative Arts, 10*(1), 74–97.

Espinós, A., Orihuela, M., et al. (1986). *El "Prado disperso". Cuadros depositados en Barcelona (I. Palacio de Pedralbes).* Madrid: Museo del Prado.

Fendelman, H. W. (1981). *Silent companions: dummy board figures of the 17th through 19th centuries: exhibition at the Rye Historical Society.* New York: The Rye Society.

Herrero Carretero, C. (2004). *Tapices de Isabel la Católica. Origen de la colección real española.* Madrid: Patrimonio Nacional.

Herrero, C., Sancho, J. L., & Martínez Cuesta, J. (1996). *Tapices y Cartones de Goya.* Madrid: Patrimonio Nacional, Lunwerg.

Lafuente Ferrari, E. (1985). *La fundamentación y los problemas de la historia del arte.* Madrid: Instituto de España.

Langmuir, E. (2006). *Imagining childhood.* New Haven: Yale University Press.

López Poza, S. (2001). El Criticón y la Tabula *Cebetis. Voz y Letra. Revista de Literatura*, XII, 2, pp. 63–84.

Maeterlinck, L. (1910). Le genre satirique, fantastique et licencieux dans la sculpture flamande et wallonne. *Les miséricordes de stalles. Art et folklore*. Paris: Jean Schemit.

Manson, M. (2001). *Jouets de toujours. De l'Antiquité à la Révolution*. Paris: Fayard.

Manson, M. (2004). La poupée et le tambour ou de l'histoire du jouet en France du XVIe au XIXe siècle. In: D. Julia, E. Becchi (Eds.). *Histoire de l'enfance en Occident*. 1. *De l'Antiquité au XVIIe siècle* (pp. 432–464). Paris: Seuil.

Manson, M. (2005). *Histoire(s) des Jouets de Noël*. Paris: Téraèdre.

Manson, M. (2007). Histoire du jouet dans l'art, approche anthropologique, 1450–1650. *Annali della Facoltà di Lettere e Filosofia, Università di Siena* (Vol. XXVI, 2005, pp. 129–164). Fiesole: Cadmo.

Marsillach, L. (1944). *Vida y tragedia de Isabel de Austria. Novela histórica de una emperatriz*. Barcelona: Hymsa.

Mateo Gómez, I. (1979). *Temas profanos en la Escultura Gótica Española. Las Sillerías de Coro*. Madrid: CSIC, Instituto Diego Velázquez.

Mestre Sancho, J. A., & Blasco Carrascosa, J. A. (1990). *Juego y deporte en la pintura de Goya*. Valencia: Generalitat Valenciana.

Pedraza, P. (1983). La Tabla de Cebes: un juguete filosófico. *Boletín del Museo e Instituto Camón Aznar, 14*, 93–110.

Pieper, J. (1999). Glück und Kontemplation. In B. Wald (Ed.). *Josef Pieper. Kulturphilosophische Schriften* (Vol. 6). Hamburg: Meiner Verlag.

Portús, J. (2006). El siglo XVII: la madurez del género. In L. Ruiz Gómez (Ed.). *El retrato español en el Prado. Del Greco a Goya* (pp. 83–89). Madrid: MNP.

Randall, L. M. C. (1966). *Images in the Margins of Gothic Manuscripts*. Berkeley: University of California Press.

Roberts, A., & Donaldson, J. (1869). *Clement of Alexandria (2) – Ante-Nicene Christian Library* (Vol. XII). Edinburgh: T. Clark.

Rosenthal, D. A. (1982). Children's Games in a Tapestry Cartoon by Goya. *Philadelphia Museum of Art Bulletin, 79*(335), 14–24.

Teijeira Pablos, M. D. (1996). *Juan de Bruselas y la sillería coral de la catedral de Zamora*. Zamora: Instituto de Estudios Zamoranos "Florian de Ocampos".

Todorov, T. (1993). *Éloge du quotidien. Essais ur la peinture hollandaise du XVIIe siècle*. Paris: A. Biro.

Vaz-Romero Trueba, O. (2011). *El artista y el juguete. Viajes al imaginario occidental, desde la Antigüedad al Romanticismo*. Barcelona: Universidad de Barcelona-Université Paris-13.

Vaz-Romero Trueba, O. (2014, June). La chambre d'enfant chez les illustrateurs espagnols: un espace magique? (1874–1986). *Strenae* 7. doi:10.4000/strenae.1260. Retrieved from http://strenae.revues.org/1260.

Vaz-Romero Trueba, O. (2016). *Puer Senex* o el niño en la pintura de El Greco y de sus contemporáneos españoles, entre Neoplatonismo y Naturalismo. In E. Almarcha Nuñez-Herrador, P. Martínez-Burgos & E. Sainz Magaña (Eds.). *El Greco en su IV Centenario: patrimonio hispánico y diálogo intercultural.* Toledo: Servicio de Publicaciones de la Universidad de Castilla-La Mancha.
Weiss, K. (1990). *Die eherne Schlange und andere kleine Prosa.* Marbach: Deutsche Schillergesellschaft ("Über die Armut im Geiste").
Wirth, J. (2008). *Les Marges à drôleries des manuscrits gothiques.* Geneva: Droz.

Authors' Biography

Dr. Oriol Vaz-Romero Trueba is Associate Professor at the Faculty of Fine Arts of the University of Barcelona. Holds a Ph.D. in Fine Arts at the University of Barcelona (UB), Spain, is European Doctor of Education Science at the University of Paris XIII, and is a painter and sculptor.

Dr. Michel Manson is Professor Emeritus at Université Paris XIII, Paris, France. He is a Doctor of Modern and Contemporary History and a member of EXPERICE—Université Paris XIII/Paris VIII.

CHAPTER 8

Communication in Moroccan Children's Toys and Play

Jean-Pierre Rossie

Introduction

The topic of communication, also called transmission, is a quite recent one in the field of ethnographic research on children. This chapter is about communication in the sphere of toy-making and play activities of Moroccan children. It mostly discusses communication between children: peers or older and younger children, but sometimes between children and adults. There is also a section on the communication between the players, their families and the ethnographer. Two important aspects, namely gender and change, are discussed in separate sections.

The games and toys described here are those of children between 3 and 15 years of age living in rural areas and popular quarters of towns. These communities largely retain their traditional ways, although there is a growing impact of modern technology on their way of life. The data on Moroccan toys and play have been collected since 1992 mostly from children belonging to the Amazigh (Berber) populations of central and southern Morocco. In many cases, the information comes directly from children but sometimes the memory of adolescents, adults, or older people was used.

J.-P. Rossie (✉)
Center for Philosophical and Humanistic Studies, Catholic University of Portugal, Braga, Portugal
e-mail: sanatoyplay@gmail.com

L. Magalhães and J. Goldstein (eds.), *Toys and Communication*,
https://doi.org/10.1057/978-1-137-59136-4_8

The toy and play cultures of North African and Saharan children have been largely neglected in scientific and popular literature. Three remarkable exceptions are Charles Béart's two volumes on toys and play in West Africa, containing some information on North Africa and the Sahara (1955), Paul Bellin's study of Saharan children's play (1963), and Fernando Pinto Cebrián's book on traditional games and toys of the Saharawi (1999). An annotated bibliography offers detailed information on these and other related publications (Rossie 2011).

In my research, the children are seen as social and cultural actors within their own communities and in their relation to the ethnographer. They are not viewed as an isolated group but as part of the community to which they belong, taking their relations with adolescents and adults and their role in developing culture and society into account. Moreover, these children's lives are embedded in specific but changing physical, material, psychological, cultural, and social environments.

The analysis of communication through Moroccan children's toys and play is based on the information written down in the e-books of the collection *Saharan and North African Toy and Play Cultures* freely available on the Internet. To facilitate finding the more detailed data, the pages in these e-books are mentioned. To offer the reader the visual support needed for this chapter, I have published on the Internet the PowerPoint presentation: 'Communication in Moroccan children's toys and play' (Rossie 2017).

My heartfelt thanks go to Khalija Jariaa (province of Tiznit) who from being a homemaker developed into a reliable ethnographic collaborator, to Boubaker Daoumani (province of Sidi Ifni) who as a primary schoolteacher offered valuable help, and to Gareth Whittaker who has helped with correcting and commenting on my English texts since 2005.

Communication Through Children's Toys and Play

In my talk during the ITRA congress in July 2014, I used the terms horizontal and vertical transmission. However, I changed this, mostly because it is impossible to say what qualifies a transmission as vertical, defined as a transmission between an adult and a child, or as horizontal, defined as a transmission between children. For example, is learning how to make a toy in a relation between an adult of 21 and a child of 14 a vertical transmission, and a transmission between a child of 14 and a child of four a horizontal transmission? If it is a question of difference in age, then this is surely doubtful, as in the first case there is a difference of 7 years and in the second case a difference of 10 years. Moreover, the process of transmission is part of the larger process of communication. Therefore, I decided to use 'communication between children' and 'communication between children and adults.'

Although the examples of communication through toys and play mentioned in this chapter belong to make-believe play, games of skill can also offer many examples. The major reason for this imbalance in examples lies in the fact that the detailed analysis of the games of skill of Moroccan children still needs to be done. The observation of children's make-believe play and of their games of skill in these regions shows a strong relation with the natural as well as the human environments in which the children grow up. Moreover, these playful activities refer to the tangible and intangible aspects of the culture and society these children live in.

Play and toy cultures are part of the communication systems by which worldviews and ways to organize life are transmitted. Playful communication extends to every domain, such as the non-verbal and verbal transmission of knowledge, beliefs, attitudes, behavior, skills, sensibilities, and emotions. However, not only is communication at stake but so is the need to relate oneself to this ongoing process. Non-verbal communication is undoubtedly primary, regarding both the temporal aspect, i.e. from the beginning of a child's life, and the quantitative and qualitative importance of what is communicated. Nevertheless, the role of verbal communication should also be stressed. Although there is as good as no information on Moroccan children's verbal communication in play, the protocols of three videos made at the beginning of 2002 in the Sidi Ifni region show that it can be an important aspect in children's play (Rossie 2014: 7–33).

In the make-believe play of Moroccan children from the end of the twentieth and the beginning of the twenty-first centuries, the link between children's play and toy making on the one hand and reality on the other is strong. In rural Morocco, it is only quite recently that some imported dolls, robots, and soft toy animals have seemed to refer to an unreal or fantastic world and, even then, the playing child may relate them to a real person or animal.

This chapter offers no room to discuss the important aspect of 'sign and meaning' in communicating through handmade toys. However, I can refer the interested reader to the chapter 'Toys, signs and meaning' in *Toys, Play, Culture and Society* (Rossie 2013a: 49–80). This chapter analyzes successively the material aspects, the technical aspects, and the cognitive and emotional aspects of North African and Saharan toys.

Discussing communication in play and toy-making activities is an excellent way to take the children's point of view into account and to illustrate children's active participation in the elaboration of their own worlds and of the cultures and communities to which they belong.

Communication Between Children

Communication through play and toy-making activities was and still is fundamental in the maturing of Moroccan children. Moreover, the basic role in this process does not belong to adults but lies in the hands of the children themselves. This playful communication occurs between older and younger children as well as between peers, whereby one should stress the importance of stable playgroups, mostly based on family and neighbour relationships. Popular quarter streets of towns and play areas in villages are real laboratories for interaction and creativity. It is there that small children, once liberated from the direct control of their mother, mix daily with children of their own age, older children and sometimes adolescents. Obtaining information, getting insight into roles and relations, learning skills and acquiring expertise through playful activities are largely dependent on the children's observation, participation, and demonstration rather than on verbal instruction.

Toddlers are quickly integrated into mixed playgroups, regularly under the control of an older girl, rarely of an older boy. Such older children certainly have an important role in the transmission of the local culture and social organization, especially the play culture and the rules of behavior between children. From the age of about 6 years, same-sex peer groups become more and more important, although the difference in age between its members can span a few years.

It is common to distinguish groups of peers from groups of older and younger children. Although I sometimes distinguish these two groups, it is difficult to label Moroccan playgroups according to this distinction, especially when the players are young and ages vary, or when older girls supervise small children. Therefore, I mention some examples based on the developed play theme rather than on the age of the participating children. By doing so, most examples show a gender differentiation separating girls' play from boys' play.

Quite a few girls' games are related to dinner-party play. These games and making toys for them, especially unfired and fired clay toys, undoubtedly transmit information and skills not only about gathering and preparing food but also on the required materials and technology, feeding habits, and rules. When girls of similar age engage in such play activities, but even more when younger girls participate, an important aspect of female life is enacted and transmitted (Rossie 2008: 173–184). The same can be said about the playful acting out of household tasks through which, among other aspects, the community's view on women's roles and duties is experienced and interpreted (Rossie 2008: 205–225). Although dinner and household play and creating toy utensils specifically belong to the play world of girls, small boys of up to about 6 years wholeheartedly participate in these activities.

Dinner-party play, household play and doll play offer many possibilities for child-to-child interaction, as these are mostly collective games. It is not exaggerated, I think, to attribute to such make-believe play, which is very popular among Moroccan girls, an important role in developing their female identity through shared playful experiences. Several examples of Moroccan dolls and doll play underpin this statement as they refer to the local roles and relationships of brides, pregnant women, mothers, grand-mothers, childless women, maids, poor women, rich women, and some female professions. A detailed example is given through the doll play and the dolls of three sisters and their mother from a village near Midelt in Central Morocco (Rossie 2005a: 124–134). More recently, Khalija Jariaa's observations and photos of children's play and toy-making activities in some Anti-Atlas villages offer a deeper insight into the doll and household play of Amazigh girls and into new play themes influenced by media, tourism, and migration (Rossie et al. 2016: 42). In Fig. 8.1, girls play with plastic dolls dressed by themselves to enact, in a village on the outskirts of Tiznit, scenes inspired by television programs relating to French tourists visiting Morocco (Magalhães and Rossie 2014).

Games referring to subsistence activities, such as working in the fields and trading, often belong to the play activities of boys (Rossie 2008: 242–267). When analyzing these games, it is sometimes astonishing how detailed the boys' analyses of male activities are, and how they enact and interpret facts and feelings related to the activities of not only their male relatives and neighbors but also of strangers. No doubt, such activities offer peers and younger boys some insight into the economic roles and professions of adult

Fig. 8.1 Girls' doll play, Anti-Atlas, 2007, photo Kh. Jariaa

men in their immediate and wider environment and, by playfully engaging in such male occupations, they foster their male identities.

Another play activity in which Moroccan boys engage more often than girls is making musical instruments and playing them. Sometimes this is a solitary occupation but, on other occasions, boys form small orchestras in which traditional rhythms, melodies, and songs, or those offered by the media, are communicated between the players and singers and possibly to an audience of other children (Rossie 2008: 285–299). Figure 8.2 shows a peer group of boys giving rhythm to their songs by hitting small and large cans. In children's games, verbal and musical expressions are mostly learned from older children in combination with non-verbal behavior like hand-clapping, rope and elastic jumping, and dancing.

In some cases, I would compare a toy-making and play activity to informal education or apprentice-like learning. For example, in 2002 in Sidi Ifni, I saw two boys of about 13 years old repairing their skateboard

Fig. 8.2 Boys' percussion orchestra, Sidi Ifni, 2005, photo J.-P. Rossie

with three wheels made of ball bearings. Then they sat down on it to run down the slope at great speed. The next day and the day after, four other boys of the same age joined them. In this playgroup, the first two boys helped their friends not only to make such a skateboard but sometimes also to steer it (Rossie 2008: 349). Another example refers to a girls' doll play about a care home for children and adults in distress. In 2009 in a village on the outskirts of Tiznit, four girls between 8 and 10 years enact a Moroccan real life event they witnessed on television. In this make-believe play, seven children between 5 and 7 years are initiated into local situations and customs about pregnancy, birth, and name giving (Rossie et al. 2016: 62–71).

Playful relations between the observed children seem mostly to be friendly and positive. Yet, each one of the three examples of doll play videotaped in 2002 in or near Sidi Ifni (Rossie 2014: 7–33) shows that Moroccan children's interactions are not always harmonious, cooperative and conflict-free. Instead, opposition, confrontation, and conflicting viewpoints and play plans do occur. These events, however, seem to be quickly resolved through strategies of disengagement, giving in, rebuking, and co-players' intervention. In any case, I did not find examples of serious physical aggression or of a player leaving the play activity in great anger. The way in which players handle opposition and conflict can be related to the way this is done in their communities, where direct confrontation between family members and friends is frowned upon.

Communication Between Children and the Adult World

Games and toys are part of the communication system by which families and communities transfer to new generations their culture and social organization. However, what takes place through these children's playful activities is not an imitation but their interpretation of the adult world, of female and male activities, of festivities and rituals. The cultural life and social organization of the family and community influences, among other things, the non-verbal and verbal components of the games, the developed themes, the time and place of their realization as well as the composition and organization of the playgroups.

As far as Moroccan children's play and toy-making activities have been observed or children and adults have given information about them, it is difficult to find any kind of tangible or intangible element from the adult world that is not represented. Numerous roles of mothers, fathers, grandparents, other family members, female and male teachers, workers, leaders, etc. are enacted in children's games. Through playful relations, children learn a lot about signs and their meanings, and the values, attitudes, and beliefs prevalent in their community. In doll play, for example, girls exchange information and feelings related to femininity, gender, weddings, pregnancy, childbirth, motherhood, household life, and funerals (Rossie 2005a: 114–190).

Feasts and rituals are strong events in Moroccan adults' lives, and children take part in them in a direct way, or indirectly through their toys and games. Through these playful activities, they participate in and communicate between them many locally valuable designs, symbols, musical, and choreographical expressions and beliefs (Rossie 2008: 306–331).

Games in which children recreate the lives of parents or other members of their family and community offer them a favorable ground for understanding reality and its symbolization, and promote their socialization in an effective way. Doll play, as with most of Moroccan children's make-believe play, presents a mirror of adult life, more especially of an idealized adult life. With few, mostly quite recent, exceptions, the dolls of girls as well as of boys represent socially valued characters, symbolizing a woman or a man in a locally enviable situation. Reference is constantly made to the positive, worthy model to which the child should conform and identify. The play events may only be related to reality in a vague way, but they can also be very precise. An example from 2009 refers to a doll play in the Anti-Atlas, where girls included in their games very specific situations. For example, the father of the bride was deceased, the husband of her sister worked in Casablanca, the husband of another sister was a vegetable vendor, and that of a third sister, being lame, remained in the village to watch the house. The themes they enacted referred to everyday life, to household activities, to trade, to disputes between brides and mothers-in-law, and to other aspects of the adult world (Rossie et al. 2016: 153–154).

On an individual level, Moroccan adults very rarely interfere in children's play except when they are too disturbed, need help, or when the situation gets seriously out of hand. When asked whether she pays attention to her children's games, a Sidi Ifni mother said that playing is certainly good for them but that it belongs to the children: 'we let them play, they can play as they like, they speak as they wish' (Rossie 2014: 16). Still, some examples mentioned in my books prove that occasionally a Moroccan mother, father, adolescent, or adult sister or brother will make a toy for their child or sibling, and that this is often related to special occasions like the *Ashura* feast.

The Ashura feast, lasting for 10 days at the beginning of the Muslim year, is a specific but not unique period, during which Moroccan adults offer toys, clothing, and sweets to children. Another period when children receive gifts is the *Mussem*, the yearly feast with its fair and popular festivities, or the *Mulud*, the anniversary day of the Prophet Muhammad. An adult might also buy a toy or other gift when visiting a village or town market. Children's anniversaries, however, often pass unnoticed, at least among popular class families (Rossie 2008: 306–331).

Anti-Atlas boys from the age of about 10 years, male adolescents, and young adults play a leading role in an exceptional parade, called *Imashar*, which takes place in Tiznit and its region during the week following the Ashura feast. Organized in groups based on living in the same quarter or village, they perform as musicians, singers, dancers, acrobats, masked figures, and large animals. The preparations for this parade start months before Ashura. In these groups, a hierarchy exists between those who have much experience and those who have little or no experience. The novice participants in particular undergo training, covering different realms of physical, musical, aesthetic, and creative performance (Rossie 2008: 317–321).

Yet, all this socialization largely takes place through the Moroccan children's own efforts, not just as a passive imitation but as an active appropriation. Moreover, one should keep in mind that this socialization is only an incidental consequence, and that the children concerned play neither to become socially adapted nor to hone their skills. They play for the fun and well-being it procures for them.

Communication and Gender

Within the world of toys and play, local viewpoints on gender differentiation are transmitted to and between children by direct as well as indirect communication. But when it comes to women's or men's roles, duties, and prerogatives in make-believe play, indirect communication prevails.

Although one can find older boys and girls playing together, separation according to gender becomes predominant in Moroccan playgroups from the age of 6 or 7 years. At that age, girls and boys mostly create their own playgroups from which the other sex is excluded. In these gendered playgroups, children experience role models and develop their identity in the company of same-sex peers. Although children's situations in toy-making and play activities largely differ according to their sex, one must be cautious with generalizing statements as there are indications that sometimes the gender cleavage can be overcome. In 2006 for example, a somewhat older Sidi Ifni girl successfully infiltrated a playgroup of four about 10-year-old boys by proposing to these boys that she should perform in their game a typical task of women—namely, cleaning their small play house (Rossie 2008: 150–151).

In 1987, in his thesis about traditional games among a Middle Atlas rural population written in French, Lahcen Oubahammou describes the difference between the play worlds of girls and boys. First of all, the Aït Ouirra girl is less favored than the Aït Ouirra boy. Though still very young, she has to dedicate herself to household tasks and so cannot enjoy childhood pleasures as much as a boy. The situation of female adolescents is even worse, as they are married from the age of 12 or 13 and enter fully into adult life with all its responsibilities and obligations (Oubahammou 1987: 126–127). This very young age of Moroccan brides seems to be exceptional, as in 1960 the average age of Moroccan rural brides was 17 years. However, the same source mentions that in 2010 more than

35,000 marriages of girls under 18, the legal marriage age, were authorized by Moroccan courts (http://www.welovebuzz.com/a-quel-age-se-marient-les-marocains, date accessed December 19, 2015).

The doll and household play of Moroccan rural girls and the toys they make for these games strongly refer to the duties and tasks of women: collecting firewood, fetching water, molding corn, baking bread, preparing oil, washing linen, spinning, weaving, and also dressing up. Especially in their wedding play, these girls exchange information, discuss roles, and experience feelings related to womanhood and men-women relations. The make-believe play of Moroccan boys refers mostly to jobs and tasks of adult men in construction work, farming, and trading (Rossie 2008) or to being police officers, soldiers, vehicle drivers, and technicians (Rossie 2013b). Playing such games together with peers clearly promotes the transmission of information and opinions based on local and media sources.

The play environment of little boys and girls is normally limited to the space adults can oversee, but the play environment of the boys widens when they become older. Older children like playing at some distance from their house, using open spaces and streets. In any case, they prefer to play where they more easily escape adult control, especially the control of those who know them well. However, a quite clear distinction must be made between boys and girls, as boys enjoy more freedom and time to play than girls do. Older boys can be found making fun kilometres away from home. Older girls on the contrary must stay near their home so that adult control is easier and they remain available for helping in the household or taking care of small children. This does not mean that girls cannot travel some distance—for example, to go to secondary school—but they are normally not allowed to do so in their free time, except to visit family members or to participate in the activities of a youth centre when they live in a town. Older girls living not far from the Atlantic Ocean may go to the beach but this happens mostly in the presence of a parent, an adult sibling, or an adult neighbour. Yet, one should remember that exceptions always exist and this is certainly the case in Moroccan villages and towns.

The fact that Moroccan primary school children often play in same-sex groups does certainly not prevent playful relations between boys and girls. In 2009, Khalija Jariaa observed an interesting example in a village 29 km from Tiznit. About 15 girls and 10 boys ranging in age from 5 to 11 benefit from the rainy weather to create clay toys. Yet, once the clay is prepared, a discussion takes place between the older boys and the older girls on the type of toys that each sex will create. The boys try to insist that the girls will create toy utensils and the boys toy animals. First, the girls refuse this diktat and say that all the children can make the toys they want. However, the big girls soon accept the boys' demand in order to restore peace, but also because they like the idea of the boys buying the girls' clay utensils and the girls the boys' clay animals. This rivalry and opposition between boys and girls in play also occurs in other play situations, as when boys in a wargame destroy some girls' small houses.

One of the videos made in 2002 in Sidi Ifni shows that older age and more experience does not always provide successful ascendancy in the communication process. For example, a barely 4-year-old boy opposes his older girl-cousin's repeated order to prepare the dolls' dinner by shouting: 'Go yourself! I am a man, not a woman!' (Magalhães and Rossie 2014: 10).

Until recently, girls' play remained more within the sphere of tradition than that of boys who quickly react to technological innovations. However, research in some Anti-Atlas villages shows that this statement must be qualified because of girls' make-believe play referring to emigrants' daughters, European female tourists, and Egyptian belly dancing. Moreover, a few girls put handmade dresses inspired by Moroccan and European fashion on their plastic dolls (Rossie et al. 2016: 101–125; 157–159).

Communication with the Ethnographer

In my publications, I have incidentally talked about my own and Khalija Jariaa's contacts with children whose toy-making and play activities were observed and photographed, with adult members of their families, or with their teachers. In this section, I will try to offer a more developed discussion of this topic.

During the first period of research in Morocco (1992–2000), I stayed consecutively for two to three months in Central Morocco, interrupted by shorter stays in Belgium. When in Morocco, I lived in Marrakech, Kenitra, Khemisset, and Midelt, places from where I explored nearby and more distant areas in order to get an idea of children's play culture in different regions and to verify information gained from analysis of the relevant bibliography and a large toy collection in the Musée de l'Homme. Therefore, contacts with Moroccan families, primary schools, and children remained occasional, although revisiting the same places and persons sometimes happened.

When trying as a foreigner to establish contact with a family from the popular class, the custom of traditional hospitality is very helpful, but in presenting oneself it was important to avoid attitudes creating suspicion of being connected to the authorities or of belonging to a more affluent milieu. Walking and riding a bicycle were adequate ways to approach local people, as was offering some transport facilities to parents with children on those occasions when I was driving a car in Morocco. Another useful way to prepare for visits to rural areas was by establishing friendly contacts with adults working in small restaurants and shops in town who, after learning about my interest in children's lives, agreed to inform their rural relatives about my coming. Primary school teachers regularly helped in establishing contact with children and their parents, served as informants, and occasionally as interpreters.

With children as well as with adults, showing interest in making toys and mentioning what I knew about toys in other Moroccan places had an ice-breaking effect. Still, when contacting parents and teachers, my great interest in children's toys and play was at least surprising to them. Trying to explain their importance in the upbringing of children, I referred to the adult's own youth, asking how he or she learned specific skills like making bread, farming, modeling clay, weaving, etc., something that often created an understanding smile and remark.

In 2001, I decided to concentrate further research in Morocco on a small region over a long period, thereby returning to the kind of micro-research I did among the Ghrib of the Tunisian Sahara (1975/1977). The results of the fieldwork among the Ghrib are integrated in the volumes of the collection *Saharan and North African Toy and Play Cultures* (Rossie 2013b).

I choose Sidi Ifni, a coastal town about 180 km south of Agadir. This has turned out to be a good choice, especially because of the collaboration that has developed with Boubaker Daoumani and Khalija Jariaa. This collaboration permits a more detailed and deeper understanding of toy and play culture in the Anti-Atlas Mountains than I have been able to offer for other Moroccan regions.

Since 2005, the largest contribution to the observation and photographing of Anti-Atlas children's toy-making and play activities has been made by Khalija Jariaa. Her role as an informant and later as an ethnographer is remarkable, and presented in the book *Make-believe play among children of the Moroccan Anti-Atlas* (Rossie et al. 2016). Finding such a local collaborator, being able to awaken her interest and train her in the field offer several advantages to a foreign researcher. One advantage is Khalija's knowledge of Tachelhit, the local Amazigh language, and Moroccan Arabic. Another advantage is linked to the fact that she is a local woman which gives her easy access to families and children, especially through her network of relatives and friends. In the village at the outskirts of Tiznit where Khalija has lived since her marriage in 2005, the girls and boys are very used to her presence in their play areas. Occupying herself with some needlework, she can observe them and discreetly take photographs with a small digital camera, and avoid disturbing their spontaneous activities. Afterwards, Khalija transmits all this information to me during working sessions and discussions in Sidi Ifni or Tiznit, at which Boubaker helps with translation and explanation.

Over this last decade, the ethics of communicating with children and their families in ethnographical research has become a crucial topic. Yet, during my studies and during most of my fieldwork, this was a completely neglected subject. Probably the first mention of ethics I wrote down reads: 'Concerning my contacts with children, the ethical rules put forward by the European Council for Scientific Research have been followed. Thus, paternal or maternal authorization has been obtained when collecting information from children or when photographing them. Certainly, it would have been difficult to do it any other way, the research being done in families or in public spaces. Still, there is one exception to this rule, namely the observations or photographs of children occasionally taken in streets or public areas in Moroccan urban centres, in which case the permission of the children themselves was only requested when taking photographs. On a few occasions, the photograph was taken from a distance, without asking the children involved for their permission. Yet, in these cases adults were present in the area and I encountered no negative reaction when photographing these children' (Rossie 2005a: 40). When I and even when Khalija were observing and photographing toy-making and play activities, children's consent was mostly expressed indirectly, by their acceptance of a discreet non-interfering presence. The children's response to an ethnographer, however, is very direct: they relate or they withdraw. Therefore, without being accepted and trusted by the children, their families, and neighborhoods, participant observation is impossible.

Showing interest in children's material culture—albeit somewhat strange, especially when an older man does so—seems to me an adequate and cautious approach. Observing children's toy making and toy use offers information not only about materials, techniques, and symbols, but also about represented roles and situations. As far as my experience shows, Moroccan children like this interest being taken in their toys and games. They feel respected and appreciated when seen as valuable informants and producers of objects and knowledge. After all, they are the ones who know and the ethnographer is the one needing to learn. Seen from this viewpoint, I decided not to follow the scientific custom of using pseudonyms or hiding locations. When asking some children what they preferred, they wanted to be correctly named and the consulted adults found this no problem. The fact that sensitive topics are seldom at stake in playful activities makes this, I think, acceptable.

Collecting toys from children, something I started to do purposively when being in the Anti-Atlas, and donating Moroccan toys to museums, can be problematic, especially when the toy is rather unique, more or less difficult or expensive to make, or belongs to a small child attached to it but whose father or mother wants to offer it to the ethnographer. In the last case, I diplomatically refused the gift. In the other two cases and when it was a rather uncommon toy, I asked the child if remaking the toy was feasible and let the child decide to hand over the observed toy or a copy. The toys donated to museums and associations are described in the series *Saharan—North African—Amazigh Children's Toy Catalogues*, available on the Internet at Academia.edu: https://independent.academia.edu/JeanPierreRossie; Scribd: https://www.scribd.com/jean_pierre_rossie; and the author's website: http://www.sanatoyplay.org.

I very rarely buy toys, and this for several reasons: they have no monetary value for the children or the adults and, being a guest or a friend of the family, one cannot pay for a gift as this could be felt as an offence. Making toys is often done in a playgroup, so how could one decide whom to remunerate or how to divide the payment? In Morocco, I integrated this collecting of toys and at the same time the hospitality received into an exchange of gifts and counter-gifts. When staying with families, my participation in this exchange was to share in expenses for children's clothes, shoes, or school fees, for household necessities, medicine, or extra food. If some feast, especially the Ashura feast, was happening or about to happen soon, a monetary gift in this context was always appreciated. When having the chance to approach pupils of a primary school class, quite soon the usual compensation became offering a financial contribution towards the costs of the end of the school-year celebration. Later on in the Anti-Atlas, girls' playgroups liked to use the pecuniary gift to buy food and lemonade for a *maaruf* or traditional happening adults organize on special occasions.

I only bought a toy when exceptionally it had become an item for tourists, or when making it necessitated buying some material in a market or shop (Rossie 2005a: 145, 2008: 294).

Of course, I would like to know in detail what Moroccan children and adults thought and think about me and my occasional and regular collaborators. Yet, this is a perilous undertaking because rules of politeness and respect create an awkward situation, making an honest response improbable. This is even more so for children, as the difference in age and status is prominent. Moreover, one must be aware if the child is or is not of the same sex as the ethnographer and that the verbal abilities of small children are limited. Yet, the clearest answer is found in a child's and a playgroup's response by accepting or rejecting being observed and communicating.

Change and Continuity in Toys and Play

Continuity of play themes, games, toy-making activities, and playgroup building is still prominent in Moroccan rural areas, although change is becoming increasingly important. Evolution has always existed but the changes have clearly accelerated since the Second World War. Only a few examples related to communication between Moroccan children are mentioned here, but the reader can find more information in Rossie (2013a: 149–182, 2008: 361–364).

The desertion of villages and the galloping urbanization in towns change the specific spaces for playful interaction by replacing open air unstructured play areas with structured spaces and by altering different aspects of the process and the content of playful communication.

The Moroccan school system strictly regulates children's playtime, sometimes proposes unfamiliar content, and imposes, especially in the case of Amazigh children, a dominant language and culture. Yet, it also creates possibilities of finding new inspiration and playmates outside the worlds of family and neighborhood.

Television programs, videos, and computer games have successively become important sources of play themes and role models for boys and girls. A good example is the great success of Pokémon starting in 2000 after a Moroccan television station broadcast an Arabic spoken version of this animation series (Rossie 2013a: 180). Action films clearly influence boys' play. The TV news is also integrated in boys' as well as girls' pretend games—for example, when they enact scenes from the Palestine war, from dramatic events as during the Casablanca flood (Rossie 2013b: 78; 93; 233), and from care homes for children and adults in distress (Rossie et al. 2016: 62–71). Media may inspire some unusual play themes, as when five about 9-year-old girls ventured to create a doll play containing belly dancing seen on TV in Egyptian films, causing them to express *h'shuma*, shame as well as hilarity (Fig. 8.3), but also concern about the expected negative reaction if adults should find out (Rossie et al. 2016: 104–106).

The massive import of cheap plastic toys made in China, nowadays found in all Moroccan markets, has created a fundamental change and sometimes stops the transmission of toy-making skills between older and younger children. Moreover, it creates a new situation whereby children develop an attitude of expecting toys as a gift from adults, an attitude that until recently was as good as non-existent in rural areas and popular quarters of towns.

Fig. 8.3 Girls' enacting belly dancing, Anti-Atlas, 2006, photo Kh. Jariaa

In make-believe play, Moroccan children also integrate and transmit new ideas, values and attitudes inspired by the way of life of tourists and of youngsters of Moroccan origin living in Europe visiting Morocco, at the same time expressing new concerns such as going on vacation and illegal migration (Rossie et al. 2016: 27; 100–125; 157–159).

The fact that some children already show at the age of ten and younger an interest in what happens outside their own community is attested by a make-believe play featuring part of a plastic doll representing a disabled girl. At the end of this game, the leading girl asks her young brother to take a photo with his handmade camera so that she can send the photo to *el hukuk el insan*, the National Human Rights Council in Morocco, to get some support (Rossie et al. 2016: 74–75).

Since I started my fieldwork on Moroccan children's toys and play in 1992, changes have become of growing importance yet they do not seem to have created great conflict between the old and the new in children's playful communications. Some games, like wedding play, bridge three dimensions of time: they refer to a girl's future as bride, build upon children's actual knowledge about weddings and man-woman relationships, and incorporate ancestral customs.

Conclusion

In play activities, children not only discuss positive aspects of their communities but also integrate in them some problematic or negative aspects of adult society. Such a situation was acted out in the doll play of three Anti-Atlas girls, when the doll representing the matron of a care home for children and adults in distress forbids a beautiful but disabled adolescent girl, represented by another doll, to leave the home as she risks being abused by men (Rossie et al. 2016: 61).

The emphasis on communication within playgroups overshadows the role of Moroccan children's personal motivation and initiative. A discussion of children's creativity I made earlier shows that, in rural social environments more oriented towards communality, children's individuality must be taken into account (Rossie 2008: 364–370, 2013a: 93–103). Research during this last decade in the Anti-Atlas will, when published, bring forward more information and possibly new insights into leadership and creativity in play.

This chapter has shown that Moroccan children are active participants in the society and culture in which they grow up and that their play and toy culture is valuable. Therefore, it should be rightly recognized as an integral part of the tangible and intangible heritage of their country. I can only hope that interest in this children's culture will grow strong enough in Morocco to avoid it surviving only in foreign museums.

References

Béart, C. (1955). *Jeux et jouets de l'Ouest Africain.* Mémoires de l'Institut Français d'Afrique Noire. Dakar: IFAN. Volume 42. 2 vols, 888 p., ill.

Bellin, P. (1963). L'enfant saharien à travers ses jeux. *Journal de la Société des Africanistes* XXXIII, 47–103, 19 ill.

Magalhães, L., & Rossie, J.-P. (2014). Children as toy makers and toy users: Television relevance in Moroccan rural child play. In *Childhood remixed: A special edition with papers drawn from the International Children and Childhoods Conference Held at UCS, July 2013* (112 p., pp. 77–85). Suffolk: Childhood Remixed Journal. Retrieved from https://www.ucs.ac.uk/Faculties-and-Centres/Faculty-of-Arts,-Business-and-Applied-Social-Science/iSEED/Childhood-Remixed-Journal-2014.pdf.

Oubahammou, L. (1987). *Ethnographie des jeux traditionnels chez les Aït Ouirra du Maroc: description et classification.* Thèse, Faculté des Sciences de l'Education. Laval: Université Laval. 147 p., ill.

Pinto Cebrián, F. (1999). *Juegos Saharauis para Jugar en la Arena. Juegos y Juguetes Tradicionales del Sáhara.* Madrid: Miraguano S.A. Ediciones. 119 p., ill.

Rossie, J.-P. (2005). *Saharan and North African Toy and Play Cultures. Children's dolls and doll play.* Foreword by Dominique Champault. Stockholm: SITREC, Royal Institute of Technology. 328 p., 163 ill.

Rossie, J.-P. (2008). *Saharan and North African toy and play cultures. Domestic life in play, games and toy.* Foreword by Gilles Brougère. Stockholm: SITREC, Royal Institute of Technology. 438 p., 410 ill.

Rossie, J.-P. (2011). *Saharan and North African toy and play cultures. Commented bibliography on play, games and toys.* Stockholm: SITREC, Royal Institute of Technology. 72 p.

Rossie, J.-P. (2013a). *Toys, play, culture and society. An anthropological approach with reference to North Africa and the Sahara.* Foreword by Brian Sutton-Smith. Stockholm: SITREC, Royal Institute of Technology. 256 p., 144 ill. Digital version of the 2005 publication.

Rossie, J.-P. (2013b). *Saharan and North African toy and play cultures. Technical activities in play, games and toys.* Foreword by Sudarshan Khanna. Braga: CEFH, Catholic University of Portugal. 360 p., 350 ill.

Rossie, J.-P. (2014). Videos on Moroccan children's play and toys available on YouTube: References and Notes, 33 p.

Rossie, J.-P. (2017). Communication in Moroccan children's toys and play. PowerPoint for Chapter 8. In L. Magalhães, & J. Goldstein (Eds.), *Toys and Communication*. UK: Palgrave Macmillan. 37 slides.

Rossie, J.-P., Jariaa, K., & Daoumani, B. (2016). *Saharan and North African toy and play cultures. Make-believe play among children of the Moroccan Anti-atlas.* Advance online publication, p. 284., 303 ill.

Author Biography

Jean-Pierre Rossie holds a Ph.D. in African Ethnology and is a founding member of the International Toy Research Association.

CHAPTER 9

Dincs as Worldviews: *Things* that Communicate a Mind

Koumudi Patil

Introduction

Things or rather *Dincs* are 'gatherings' that are assembled to draw people towards the concerns they stand for. Heidegger (1971: 174–182) ushered the etymology of '*dincs*' into contemporary discourse from gatherings of free Nordic and Germanic men discussing and negotiating concerns of their societies. These pre-Christian *things* were assemblies, rituals, and places where disputes were solved and political decisions taken (Telier et al. 2011: 1). Thus, the etymology of the English word '*thing*' reveals its trajectory from an assembly dealing with 'matters of concern' common to the community, to meaning an object—'an entity of matter,' often associated with 'matters of fact.'

Hence, a *thing* is, in one sense, an object out there and, in another sense, an issue (Latour 2004: 233). As a matter of concern, '*thinging*' gathers human beings, and *things* are events in the life of a community that play a central role in their common experience (Heidegger 1971: 174–182). *Things* around us constitute the everyday fabric for experiencing and making sense of the world (Telier et al. 2011: 52). Such social grounding of *things* questions how the social is gathered, ordered, and

K. Patil (✉)
Indian Institute of Technology, Powai, India
e-mail: kppatil@iitk.ac.in

L. Magalhães and J. Goldstein (eds.), *Toys and Communication*,
https://doi.org/10.1057/978-1-137-59136-4_9

structured through spatiality and materiality of the *thing*. How is this sense of the world communicated by a *thing*?

According to Brown (1998: 935), if the history *of things* can be understood as its 'social life' through diverse cultural fields, then the history *in things* might be understood as the crystallization of the anxieties and aspirations that linger in the material object. To illuminate the historical circulation of *things*, we have to follow the *things* themselves, for their meanings are inscribed in their forms, their usage, their trajectories (Appadurai 1988: 5). *Things* are embodiments of human thoughts. In the sense that *things* are a product of a human mind, the analysis of their trajectories can interpret the human negotiations that breathe life into *things*.

As an object, a *thing* resonates well with a community that favors bricolage and performativity. *Objeu* from 'object' and the French word *jeu* ('play') (Cornell 1993 as cited in Telier et al. 2011: 53) summarizes the primary occupation of its user—*homo ludens*—to play. Like *thinging*, playing is a self-supporting and independent pursuit. It is an intrinsic activity that exists for its own sake, outside of ordinary life (Huizinga 1955: 9). Accordingly, *jeu* or 'play' summarizes the *thing* as a ritual, rite, or recreation. Players gather around a *thing* to celebrate rituals, display their exotic value, or for mere ludic entertainment. As play manifests culture (Huizinga 1955: 1–28), *things* as toys or *play-things* manifest a sense of the world, perhaps a worldview.

In this paper, I attempt to comprehend the *thing* neither as a self-supporting, independent object, nor as something made in a material or as a consequence of the four-fold. Instead, I cogitate on its quintessential property to gather, negotiate, and communicate matters of concern.

The Social Communication of the Mind

Even before we dwell on the kind of concerns, and to whom they are addressed in a gathering, we may primarily want to explore the communicative value of a *play-thing*, if it has any at all. How does the *play-thing* represent or communicate matters of concern emerging from the community worldview for it to be realized as a *thing*, and not as an object? Any theory of communication will profess that what is communicated will primarily have to be encoded in the emissary—a *thing* or a human. What is not encoded cannot be decoded by the recipient. Of course, in the process some information may be lost or an aberration may occur while decoding. A derivative of this would postulate that *things* are embodiments of human thoughts, which can be, and often are, communicated to the recipient. They reflect the use, the user, and the useful in a culture. As Eco (1976: 21) would assert, the communicative phenomena include the production and employment of objects used for transforming the relationship between man and nature. No *thing* works unless it embodies ideas that are held in common by the people for whom it is intended. It is only by showing the ways in which ideas about the world and social relations are turned into a *thing*, that we can properly comprehend the nature of a *thing*, in this case a *play-thing* (Forty 1986: 245). Thus, a *play-thing* can act as a piece of epistemological evidence that has not only witnessed, but also physically manifested, the mind of a community. If so, how do we understand what the *thing* communicates or embodies?

To this end, Dundes (1993: 88–91), in his seminal essay on 'Pecking Chickens: A Folk Toy as a Source for the Study of Worldview,' examined pecking toys from India, Italy, Poland, Spain, Russia, and the United States, deducing consumption patterns of food as well as space in the cultures to which they belong. Food and space are matters of concern often raised in negotiation and debate in any worldview, further reflecting in its microsphere—a toy. Dundes observed that Polish chickens have very little space between the individual heads and peck from a single communal bowl, which is virtually empty. On the other hand, the use of real kernels of corn prompted him to argue that only a country with abundant food supply would waste food on decorating a toy. Upon observing the Italian toy with an egg, he postulated its resemblance to the mother-child configuration in Italian culture.

However, what intrigued me most, was Dundes' interpretation of the Banarasi pecking toy. (Refer to Plate 1.) He hinted that the eight chickens pecking on the plate indicated the population explosion of the sub-continent as well as the corresponding difficulties in providing food security. In fact, art or cosmology may be considered as a worthy substitute for food in the Indian culture, because the chickens are pecking not food, but attractive scenery!

Though Dundes frequently cautions against making speculations from a few texts or exemplars, he nonetheless attempted to explicate the implicit social construct of the world as gathered in the toys by the toymakers. However, as an ethnographer engaged in a close study of the Banarasi toymakers in Varanasi, India, since 2008, I find the observations made on the Banarasi pecking toys inadequate to explain the rich complexity of the Banarasi mind and expression. With an increased sample size and a more rigorous methodology, maybe we can attempt to draw a more complete picture. Towards this goal, this section delves into a systemic method for revealing views gathered and communicated by the Banarasi community in Banarasi toys or *play-things.*

> For this study, 150 *play-things* were collected through direct purchase from the Banarasi shops, antique warehouses, or from toymakers themselves, besides taking images of *play-things* archived in the Banaras Hindu University, Warwickshire Museum Service, V&A Museum, and the Craft Museum, New Delhi, ranging in time from 1900 to 2011. From this pool, a detailed feature analysis of 40 *play-things* was conducted, while the rest of the toys were used as precedents to understand the evolution of the selected samples. However, to limit the scope of this paper, analysis of only the Banarasi pecking toys and their variations are discussed. (Refer to Fig. 9.1.)
>
> The origin of the pecking toy in Banaras is difficult to establish, due to the paucity of archival records, though it is established that it was assimilated from a client sample of an export order. Over time, the pecking toys as well as their variations have acquired indigenous names, such as '*chugti chidiya*'—a literal translation of the English 'pecking bird.' Its variations in Fig. 9.1 are called '*hilanta,*' from the verb *hilana,* which means 'to move.' Both tail and head follow an up-down trajectory. It is often placed on a wooden base. If this base is triangular, it has three birds, if octagonal eight birds, or, if the base is a single rectangle, it has one bird or a pair of birds.

The method of elucidating a *play-thing* as a microsphere entailed two stages—

(1) Feature analysis: features of every *play-thing* were segregated, coded and analyzed separately. Feature, here, is a distinctive element of a

Fig. 9.1 Banarasi pecking toys

play-thing that allows itself to be modified in varying degrees, besides sometimes relinquishing itself completely in successive iterations. Its external representation facilitates its coding through any means—visual, written or verbal. Such features related to the artefact (visual, shape, and size), its production (manufacturing, mechanism, and material) and consumption (purpose and connotation) were studied meticulously to arrive at patterns of frequently occurring features.

(2) Thereupon, feature analysis was used to deduce a pattern through a rigorous to-and-fro iterative process of comparison of similarities and differences between features. A systematic two-step approach with three sub-steps was followed to maintain rigor and decode the views gathered and manifested in the *play-thing*.

Step 1—The analysis of a single feature of the instance was written in the form of a statement.

The simple statement included a subject and a predicate, where the subject asserted a particular attribute in its predicate. For instance: the *pecking birds* (subject) have *round eyes* (predicate).

Step 2—Every feature of the 39 other selected toys was corroborated with Statement 1 to see:

(a) Whether it asserts Statement 1.
(b) Whether it denies Statement 1.
(c) Whether it is a new statement: that is, both the subject and the predicate are different from Statement 1.

Step 2a—If the new subject asserted the predicate of the statement, it was added to it.

Therefore, parrots and all other unidentified birds were subsequently added to Statement 1. Thereupon, the revised statement read: the *pecking birds, parrots, and all other unidentified birds* (subject) have *round eyes* (predicate).

Step 2b—The subject was rejected if many similar subjects negated the attribute. However, rare aberrations were considered an exception. For instance, the peacock (subject) has kamalnayani (lotus-like) eyes (predicate). This became an exception to Statement 1.

Step 2c—If the new feature had a different subject and attribute from Statement 1, then the process was restarted from Step 1 with a new statement.

Step 3—The statements so derived were coded together as conventions, canons, and beliefs.

Conventions, canons, and beliefs are as much about making a *play-thing* as they are a part of the social fabric. They are semantic units that are not restricted to the *play-thing* alone, but instead form part of the larger gathering of ceremonies, rituals, festivals, and everyday life, thus manifesting the world not just of the *play-thing*, but also that of its players.

Conventions are dicta that form an appropriate knowhow which habitually shapes decisions of making, using, and evaluating a *play-thing*, based on preceding instances, experiences, or conditions. Therefore, they are simple statements of assertions or denials of a particular attribute of a *play-thing*. For instance, all birds have round eyes.

Canons are community-practiced norms that refer to the assembly of typical representations or frequently repeated visible motifs, shapes, or patterns shared among all *play-things*, as well as toymakers, due to shared preceding instances, experiences, or conditions. Some canons, such as the proportions of temple architecture, are specifically mentioned in older texts, while others are passed orally or by emulating the master. For instance, pecking toys will stand on a wooden base.

Beliefs are values and notions that govern the relationship between players and *play-things*. These may be expressed through social and religious values, taboos, or superstitions, based on preceding instances, experiences, or conditions. Some beliefs are mentioned in scriptures and texts, while others are orally transferred, or observed at home or in society. For instance, the Banarasi community believes that the parrot is one of the favourite animals of the revered *Thakurji*.

Every statement was corroborated with all 40 selected toys unless it was strongly refuted by several other instances. The iterative and systematic comparison between features refined the similarities and made the aberrations apparent, as shown in Table 9.1.

This was a conscious attempt to explicate the embodied semantics communicated by *play-things*, and not an 'accidental,' 'chance,' or 'subjective' reading of *things*.[1] Together, these canons, conventions, and beliefs not only form standards, rules, and guidelines for assembling a toy, but also for viewing perceptions and gathering players around a Banarasi *play-thing*.

Moreover, it is based on these community-held semantics that a toy is understood as both object and *thing*. As an object, the toy is a matter of fact characterized by various attributes such as 'wooden, ludic, cheap, and amusing.' But the toy-object metamorphoses into a *play-thing* as and when

Table 9.1 Similarities and aberrations between toys

Convention 1	Birds are represented not with *kamalnayani*, but with round eyes. Proper eyes of *play-things* are *kamalnayani*
Canon 1	
a	*Kamalnayani* eyes are often a big white boat-shaped form, with a black full- or half-circle representing a pupil, spotted with a highlight sometimes. Eyebrow is marked in a single curved black stroke above the eye
b	Eyes of an elephant and a peacock have an extra swirling line similar to the body decoration of live animals during festivals
c	Animals will often have *kamalnayani* eyes
d	Expressions of a *play-thing* are stoic or neutral, without any cues indicating its emotional state. Smiling or drooping lips, tearful or squinting eyes, contorted eyebrows, and other similar features indicating the mood are absent
Belief 1	Banarasi community considers *kamalnayani* eyes beautiful, as do the *Shastras*—Hindu texts
Convention 2	A red patch is traditionally added onto the green wings of the *sugga* or the parrot
Canon 2	
a	Often, a complementary or a highly contrasting color is used to make a swift small brushstroke on the wing of a bird
b	Cows and horses do not have a blotch but a repetition of circles
c	All parrots and some peacocks have a red blotch on their wings
Belief 2	
a	Vermillion is smeared on auspicious images and is also a mark of fertility on a woman's forehead
b	A black blotch on the body wards off the evil eye
c	Toymakers consider a contrasting color patch aesthetic
Exception 2	Unidentified birds and pecking birds may not have a blotch on their wings
Convention 3	The Banarasi identity of the *play-thing* lies in its visual representation through painting, and not in its shape
Canon 3	
a	'*Chatak*' colors are often used to visually represent a *play-thing*. Such colors are bright, with high contrast value. Traditionally, golden yellow, rusty red, emerald green, creamy white, dull black and earthy brown were used in *play-things*. Yellow and white substitute for gold/ silver in jewelry
b	A flat base colour is applied on most *play-things* before painting motifs which often have a hard edge or border
c	Often, blank spaces on a *play-thing* will be filled up with dots, round body patches, border, ornamentation, or, recently, texture
d	Small detailed features of a body will be painted and not carved, such as eyes, mustache, lips, toenails, ornamentation, etc. This holds true for human figures that share similar body parts but are differentiated through painted details only

(continued)

Table 9.1 (continued)

e	Straight lines or dots of white, black, and yellow are interspersed with red to represent ornaments
f	Shape will be more stylized or abstract than painted features in *khilonas* for ludic purpose alone
g	A base or a platform is often made for *play-things,* which, if triangular, has three birds, while octagonal has eight birds, and single rectangle has one bird or a pair of birds
h	Shape is willfully borrowed from other *play-things* or non-Banarasi toys, but not their visual features
Belief 3	
a	Discordant features will sometimes become part of the Banarasi identity in the absence of a suitable replacement or if frequently emulated in the community
Convention 4	The ideal image consists of a complete body with stylized naturalism
Canon 4	
a	Body features are stylized, but not exaggerated like a caricature
b	Diagonal or oblique axis in the body posture is rare
Belief 5	
a	A decapitated body is considered inauspicious, unless it is part of a mythological character or story
Convention 5	Largely, *play-things* play an enculturative role in the community
Canon 5	
a	Characters and stories are often borrowed from mythology and other religious texts and epics
b	The elephant, tiger, peacock, parrot, and cow are the five favorite animals of *Thakurji*
c	Familiar characters from *Panchatantra* and folk stories are not included in the *play-things*
d	'Plain' *play-things* displaying only wooden texture are made for non-Banarasi consumers alone
Belief 5	
a	Feeding birds on the *ghats* and owning a parrot, keeping its wooden replica or painting its picture on the wall, is considered pious
b	*Play-things* teach children values and culturally upheld morals
c	Unfamiliar characters, or even those that break a taboo, can be made for non-Banarasi consumers. But these *play-things* are not sold in the local market. For instance, the owl is considered inauspicious, but is nevertheless made and sold to Bengali pilgrims

(continued)

Table 9.1 (continued)

Convention 6	*Play-things* are made together to facilitate production and efficiency
Canon 6	
a	Body parts are made interchangeable to increase design variations and meet growing market demand. Hands, feet and, at times, even faces are interchanged, while the identity is revealed through painted features
b	Manufacturing techniques can be interchanged for different parts of an artefact, but the whole artefact cannot be made in an unpracticed or unfamiliar technique
c	Willful innovation of a new mechanism without the aid of a preceding technique will not be attempted
d	Shape is willfully borrowed from other *play-things* or non-Banarasi toys, but visual features are not
e	Toymakers add characters or events in two ways: (1) They modify an existing *play-thing* to make a new one, such as a tiger's body converted into a dog (2) They borrow visual cues from popular and easily available illustrations in calendars, advertisements, and posters
Belief 6	
a	Broken *play-things*, especially those related to a religious story, are often offered to the river
b	The toymakers—*Vishwakarmas* and *Prajapatis*—claim to be the direct descendants of the gods *Vishwakarma* or *Prajapati*
Convention 7	Toys are a miniaturized world of the Banarasi adult view
Canon 7	
a	All characters in a single *play-thing* will share the same size, irrespective of their real dimensional difference
b	*Play-things* will generally have a size that can be easily handled in the palm, especially that of a child
c	Colors represent the real world unless stated otherwise in the mythology Therefore, a parrot is green, a cow is white, and a tiger is yellow, while mythical/religious characters like *Krishna* are blue, *Saraswati* white and so on
Belief 7	
Convention 8	Status of a character is represented through props held by them
Canon 8	
a	Royalty in animals like the King's elephant is depicted with body decoration and *chadar* or a cover on the back with a checkered design and borders
b	Saddlery is used to depict the status of animals as vehicles
Belief 8	
a	Despite democracy, the Banarasi community still respects and believes in the *Kashi Naresh*, the erstwhile monarch. He continues to attend ceremonies and festivals on a lavishly decorated elephant

players gather around it to negotiate certain matters of concern related to views, players, and assembly. This brings us to our other question: what kinds of concerns are communicated by a *play-thing*, and to whom are they addressed in a gathering?

Matters of Concern: Toy-Object to *Play-Thing*

Alan Dundes (1993: 83–91) examined the pecking toys to unravel how a toy may plausibly be regarded as a microcosm of a culture in which it is made. He argued that worldview permeates all aspects of a given culture. Consequently, the pattern of the whole is to be found even in the whole's smallest part—that is a toy, for toys are almost always miniatures, or 'microspheres,' of the real or adult world. Worldview is a powerful expression of how a culture *sees* the *world*, and makes it visible. A new material culture redefines both what it is to see, and what there is to see (Alpers as cited by Latour 1986: 9–10). Research has not always sought to elucidate toys from this perspective,instead preferring to merely catalog toy specimens or to detail their manufacturing and mechanism. If so, can a *play-thing* gather and negotiate the worldview of a community?

Gathering of Views

In Banaras, stories and story-telling are an important part of life, so much so that Jeeravati Devi, a local resident, elucidates that *play-things* are representations of stories. These stories are read from various ancient texts such as *puranas* and epics, or popularly recited at home or in rituals (*paath, pravachan*, or *katha-kathan*) by priests (*pandits* or scholars). Stories also unfold in the performances staged during the various fairs of Banaras. Dongerkey (1954: 64–65) says that a child playing with *things* representing religious personalities or characters learns to handle them with great care because of the reverence and respect which these *things* evoke in his mind. The child's destructive tendency can be checked by religious subject matter. That is, through *thinging*, or playing, the Banarasi community addresses matters of concern related to societal roles, norms, and values. The *play-thing* embodies an enculturative mechanism (Schwartzman 1978). Parents in Banaras believe that *play-things* not only teach beautiful stories, but also nurture cultural ethics and morals in children.

Enculturation is a matter of concern in Banaras, and therefore toy-makers refrain from portraying social vices in *play-things* that may corrupt a young mind. *Asabhya* is a Banarasi notion that denotes indecent or uncultured representations, unfit in the worldview. For instance, drinking is held in contempt, which makes the drunkard an unacceptable character in *play-things*. Similarly, a Barbie doll is considered an inappropriate object for a young girl in Banaras.

Barbie is uncouth. She allures and distracts a young mind. The best doll for a young girl should look like Radha.

Extract 1: Jeeravati Devi, housewife residing in Gadolia, Banaras.

Radha is the childhood friend and lover of Lord Krishna. She is depicted wearing a saree revealing no body part, a veil covering her head, *bindi* on the forehead, an ample figure without exaggerated curves, holding a lotus in one hand, and with no sign of make-up. Unlike Barbie, her representation can hardly allure a young child into 'adult fantasies.' A *play-thing* gathers such cues of acceptable behavior either through societal norms or, in the case of deities, from holy texts. Consequently, Vishnu and Krishna have a blue body, the eyes of all living beings are *kamalnayani* or lotus-like, Brahma is shown with three heads, while Ravana has ten. This also holds true for the *vahanas,* or the celestial vehicles assigned to every deity to pursue their ethereal excursions. The combination of the vehicles, often an animal or a bird, with the deity is unalterable: Lord Ganesha on the rat, Lord Shiva on the *nandi*, or Lord Kartikeya riding the peacock.

Ajit Kumar Vishwakarma, a toymaker, in a creative spree changed the vehicle of Lord Ganesha from a rat to a peacock in a *play-thing*. In the ensuing gathering around the *thing*, Ajit's deviation was quickly pointed out by local consumers and foreigners alike, who boycotted its purchase. Eventually, he was forced to withdraw it from the market (refer to Plate 5.3). Here, the players negotiated the meaning of the *play-thing* in the gathering, ultimately reaching a consensus. This exemplifies Latour's (2008: 4) statement that *things* are a space for negotiations. They are complex assemblies of contradictory issues.

Gathering of Players

Players are drawn towards *play-things* not only for amusement, but also to enact stories, hold ceremonies and practice rituals, the collective concern being the embodiment of an enculturative mechanism. Children as well as adults assemble and enact mythical and religious stories through *things* during various festivals. Here, children may rely equally on stories as a medium of play, as well as a stimulant for their imagination (Roopnarine et al. 1994: 23). Heroic stories of Lord Krishna's life-time are enacted by a troupe of *play-things* in tableaus to celebrate his birth during the festival of *Janmashtami*. *Play-things* representing statues of deities are choreographed in the tableaus by children and adults alike. Here, *thinging* or *playing* is an activity not just between child and object, or child and child, but also child and adult, where the child and adult are both—participant and spectator.

Similarly, during *Ramleela*, various masks of gods and goddesses are worn by children, who play around on the streets enacting stories of the triumph of good over evil with bows, arrows, and swords. Here, the story is played by actors rather than with figurines of deities, unlike *Janmasthmi* tableaus.

Of course, not all *play-things* are a part of festivals, ceremonies, and rituals. Children do gather around *play-things*, and also gather *play-things*, purely for amusement as well. These *play-things* are not similar to the ritual statues used in the *Banarasi* festivals and ceremonies. *Play-things* that are made for ludic purposes alone are often miniaturized simplified representations of reality, and exhibit bright colors. In this sense, *play-things* signify an ideal adult life which engage children in pretend marriages, processions, and other celebrations, with objects such as elephants, kings, and queens (Roopnarine et al. 1994: 19–24), as well as helicopters, *lattu* (spinning tops), *chugti chidiya* (pecking toys), *hilanta* (a single pecking animal), *sukhi parivar* (Banarasi nested dolls), animals, and birds, amongst many others (refer to Plate 3.3). Of course, such *play-things* can also exemplify play for its own sake.

However, their ludic character does not liberate them from the concerns of enculturation altogether. The *hilanta* toy, an indigenously developed two-way pecking toy, traditionally had a variety of six characters only—peacock, tiger, cow, parrot, elephant, and horse. The gathering of these particular animals is not a serendipitous encounter, as they are the favourite animals of Lord Krishna or Thakurji. Similarly, the Russian nested doll made extensively by the lathe turners includes Banarasi characters like a veiled woman with a *bindi*, a priest, or even a tiger and an elephant. Nonetheless, non-enculturative animals like penguins and cats or inauspicious ones like owls are skillfully made for the urban or western consumer only.

Not only do players gather around *play-things*, but the *things* also gather around players.

In wedding celebrations as well as baby showers, *play-things* of all kinds and sizes are gathered into the bride's lap. This gift, called *bharai*, from the parents, literally refers to filling the bride's lap with an auspicious offering or gift. A set of *chusni* (pacifiers) and *jhunjhuna* (rattles), amongst other things, are included in this ritual as symbols of fertility. *Gudda*—a male doll —is specially offered, signifying the birth of the first son. Here, *play-things* are collected for and by the adult players. Bereft of its function, the toy becomes part of a 'collection,' as in that of a collector who calls it 'a beautiful object' rather than specifying it as a 'beautiful toy' (Baudrillard 1994: 8). Thus, the gathering of *play-things* becomes a collection that signifies not need, but rather desire.

Gathering to Assemble *Things*

The adult players gather either as spectators or collectors of *things*, or as participants to assemble a *thing*. The independent parts of a *play-thing* are objects or matters of fact that are either not meaningful in themselves or do not sufficiently corroborate the sensibility of the players. It is only when they are assembled together that a *thing*—a matter of concern–emerges.

During the multiple ethnographic visits I undertook inside the Banarasi community, I observed the toymakers gathering informally, on the basis of diverse skill sets, proximity and familiarity, around a *play-thing*. Toymakers organize themselves according to skill sets, such as carving, cutting, assembling flat templates, turning wood on the lathe machine, or painting. Every toymaker produces only a part of the toy based upon either a precedent or the preceding work of the previous toymaker/s in the sequence, before passing it on to another. The latter is expected to continue shaping a coherent *thing* according to his skill, while also preserving or complementing the changes made by the former, sometimes without any or few forwarding instructions (Patil and Athavankar 2012: 38; Patil 2015: 2). (See Fig. 9.2.)

Often, such leaderless practices are reflected in the freedom to evolve permutations and combinations of differently skilled members for every *thing*. In addition to skills, even parts of toys are shared with each other in different combinations to assemble a concerned *thing*. The hands, feet and, at times, even faces are interchanged, where all figures are similarly proportioned. In Fig. 9.3, the peacock, duck, and crow all share the same body parts—wing, feet, base, square weight—as well as the mechanism and size, with minimal changes. The beak of the peacock is elongated and sharpened for the crow, while it is widened and rounded for the duck. Here, combinations and permutations of existing parts, assisted by visual

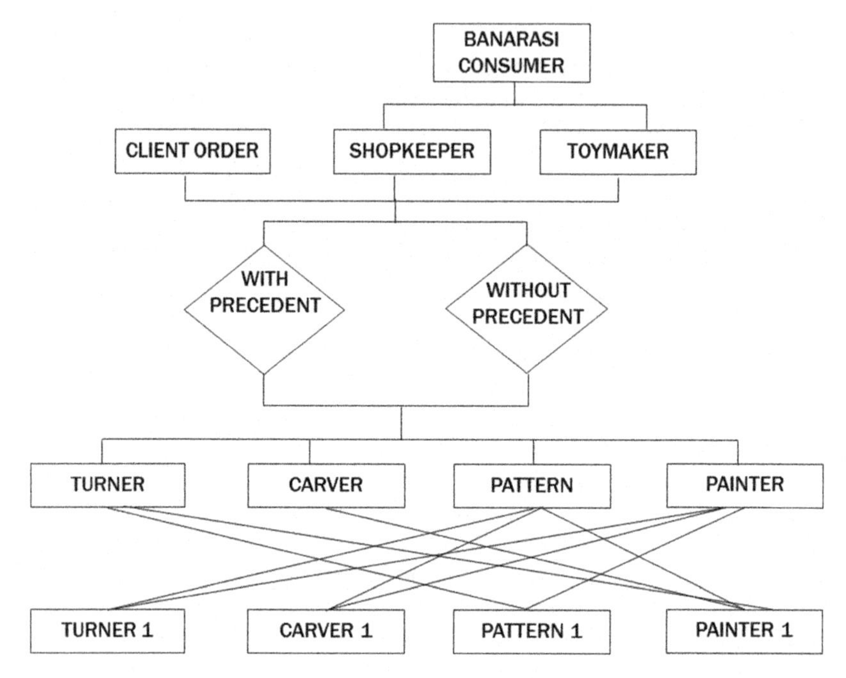

Fig. 9.2 Leaderless practice of shaping coherent *Things*

painted cues—body colour, shape of the eye, feathers, and claws—have changed the semantics of the *things* (Patil 2015: 8).

Therefore, a *play-thing* is not made by an individual toymaker but by the cumulative work of a gathering of concerned community members. Here, the mastery does not reside in any one member of the team but in the 'gathering of the community of practice of which he, (in this case the toymaker) is a part' (Lave and Wenger 1991: 4). These gatherings of toymakers, sellers, and consumers drawn around the *play-thing* negotiate the matters of concern essential to the worldview of the Banarasi community. Coomarswamy (1977: 80) describes these gatherings succinctly. He says it is the business of the toymaker to know how *things* ought to be made and to be able to do so accordingly, just as it is the business of the seller to know what *things* ought to be made, and of the consumer to know what *things* have been well and truly made and to be able to use them appropriately.

Fig. 9.3 Combination of skilled parts for Peacock, Duck and Crow toys

What concerns us most here is the ability of *things* to gather views and then communicate them as concerns to players while playing, and also to non-players, like me, while analyzing. Some concerns remain contested, which are resolved only via negotiations over a period of time, while some remain unresolved, and others get accepted. In this sense, *play-things* communicate not only what they are, but also the mind of the players.

These gatherings of community members in rituals and ceremonies as well as for assembling and negotiating the appropriate embodiment of canons, conventions, and beliefs in Banarasi toys, in rituals, ceremonies or otherwise, through seasonal and local craftsmanship in materials found close by, broaden the definition of Banarasi toys from a ludic expression alone to a cultural representation. That is, the Banarasi gatherings transform a matter of fact into a matter of concern.

Note

1. An expanded version of this table, consisting of 23 conventions, 61 canons, and 41 beliefs is a part of my doctoral thesis that included the entire collection of 40 toys, rather than only the pecking toys.

References

Appadurai, A. (1988). *The social life of things: Commodities in cultural perspective.* Cambridge: Cambridge University Press.

Baudrillard, J. (1994). The system of collecting. In J. Elsener & R. Cardinal (Eds.), *The cultures of collecting* (pp. 8–38). London: Reaktion Publishers.

Brown, B. (1998). How to do things with things (a toy story). *Critical Inquiry, 24*(4), 935–964.

Coomarswamy. (1977). The part of art in Indian life. In R. Lipsey (Ed.), *Coomarswamy: Selected papers* (pp. 80). New Jersey: Princeton University Press.

Cornell, P. (1993). *Saker. Om tingens synlighet.* (Quotes translated by Per Linde.) Hedemora: Gidlunds Förlag.

Eco, U. (1976). *A theory of semiotics.* Bloomington: Indiana University Press.

Forty, A. (1986). *Objects of Desire.* New York: Pantheon Books.

Heidegger, M. (1971). *Poetry, Language, Thought.* New York: Harper Perennial Modern Thought (reissued in 2013).

Huizinga, J. (1955). *Homo ludens.* Boston: The Beacon Press.

Latour, B. (1986). Visualization and cognition: Thinking with eyes and hands. *Knowledge and society studies in the sociology of culture past and present* (vol. 6, pp. 1–40). Greenwich: Jai Press.

Latour, B. (2004). Why has critique run out of steam? From matters of fact to matters of concern. *Critical inquiry, 30*(2), 225–248.

Latour, B. (2008). *A Cautious Prometheus? A few steps toward a philosophy of design (with special attention to Peter Sloterdijk).* Falmouth, Cornwall, Annual International Conference of the Design History Society.

Lave, J., & Wenger, E. (1991). *Situated learning. legitimate peripheral participation.* Cambridge: University of Cambridge Press.

Patil, K. (2015). Designing without a designer: A case study of decision making in banarasi toys. *The International Journal of Designed Objects, 9*(3), 1–17.

Patil, K., & Athavankar, U. (2012). *Un-authored artifacts: It takes a whole community to make a khilona* (pp. 37–45). Melbourne: Australia, University of South Denmark.

Roopnarine, J., Hossain, Z., Gill, P., & Brophy, H. (1994). Play in East Indian context. *Children's play in diverse cultures* (pp. 9–31). Albany: State University of New York.

Schwartzman, B. (1978). *Transformations: The anthropology of children's play.* New York: Plenum Press.

Telier, A., et al. (2011). *Design things.* Massachusetts: The MIT Press.

Author Biography

Koumudi Patil is Assistant Professor in the Department of Humanities and Social Sciences and Design Program at the Indian Institute of Technology Kanpur, India.

CHAPTER 10

Holocaust War Games: Playing with Genocide

Suzanne Seriff

Cultural historians and game developers have long used toy and game studies to document a society's changing values over time (Whitehill 1999: 116). As toy historian Pamela B. Nelson has written,

> Toys, like other artifacts of material culture, can tell us a great deal about changing cultural attitudes and values, and about the exercise of power in society. Mass-produced toys are especially revealing because their designers, concerned with marketability, intentionally try to appeal to dominant attitudes and values. Since the toys reflect the attitudes of the dominant group, they have helped legitimate the ideas, values, and experiences of that group while discrediting the ideas, values, and experiences of others, helping the favored group define itself as superior and justify its dominance (P. B. Nelson 1990: 10).

One genre of toys and games that has been the subject of renewed interest and attention of late includes games of war and propaganda. War games, scholars suggest—like other popular culture items such as magazine and

S. Seriff (✉)
Department of Anthropology, University of Texas at Austin,
Austin, TX, USA
e-mail: seriff@aol.com

L. Magalhães and J. Goldstein (eds.), *Toys and Communication*,
https://doi.org/10.1057/978-1-137-59136-4_10

matchstick covers, cartoons, and television shows—are some of the most common, and most powerful, means for transmitting and fomenting propagandistic messages during times of war or national unrest.

Communication scholars Garth Jowett and Victoria O'Donnell (2006: 7) have provided a concise, workable definition of the term propaganda: 'the deliberate, systematic attempt to shape perceptions, manipulate cognitions, and direct behavior to achieve a response that furthers the desired intent of the propagandist.'

The mechanism for such communication centers around 'the controlled transmission of one-sided messages via mass and direct media channels, including through children's games, toys, and literature' (R. A. Nelson 1996: 232–233).

In times of war or national threat, propagandistic media are used in a very specific way—to dehumanize and create hatred toward a supposed enemy, either internal or external, by creating a false image in the mind. This can be done by using derogatory or racist terms or names, making allegations of enemy atrocities, demonizing the 'enemy,' or making use of racist themes in rules of play.[1] Most propaganda wars—which can be effective long beyond or before an official call to arm—require the home population to feel the enemy has inflicted an injustice, which may be fictitious or may be based on facts. The home population must also decide that the cause of their nation is just.

In this chapter, I want to explore two games that draw for their subject matter on the demonization, persecution, and ultimate annihilation of Europe's Jewish population by the Nazi regime during the Holocaust of Germany's Third Reich. One game was sold to children as an inexpensive commercial item on the eve of Germany's escalating violence against its Jewish population in the 1930s; the other was created 65 years later as a 'playful' piece of art designed to confront the evils of genocidal destruction in a postmodern age. While the first would easily fit into the category of propagandistic popular culture in a time of impending war, the second, I will argue, raises important questions about the ways in which societies perpetuate historical discourses of dominance and discrimination against targeted 'demonized' populations, even in times of relative peace. Drawing on critical theorists Michel Foucault's (1980) and Stuart Hall's (1997) insistence on understanding the role of power and authority in guiding a society's shared construction of meanings not only within but across historical periods, I focus especially on the role that games and toys play in perpetuating a dominant visual discourse about European Jews as a 'legitimate' target for exploitation and even annihilation, in acts of both play and war.

While each of these two 'games' has been extensively analyzed within its own historical and presentational contexts, my interest in exploring the two together is to attempt to understand more fully the ways in which children's games and child's play may be seen not just as a *reflection* of our society's most troubled 'hot spots,' but as continual shapers or instigators of such hot spots over time. As trauma historian Robert S. Leventhal has written, 'The way in which a culture organizes, "disciplines," and reads a certain event is an excellent way to find out about that culture's troubled areas or "hot spots"' (Leventhal 1995).

If the Nazi-era game can so easily be understood as an example of propagandistic hate speech, can a late-twentieth-century Holocaust game-as-art be so easily excused from such a label—and its potentially violent real-life effects? And what do they both hold in common in terms of the seductive power of children's games and toys to trivialize, normalize, or even incite xenophobic violence under the guise of 'harmless' play?

Whether or not these games were ever widely played (or intended to be played), there is no question that the player (real or imagined) is intended to be in the position of perpetrator of targeted, hate-based crimes—and ultimately mass murder—against the Jewish population of Europe. Both provide an opportunity for children, or adults, to vicariously 'play' at the most sadistic aspects of history and human nature. What role, then, do such games and/or their artistic doubles actually play in fomenting racial, ethnic or religious hatred and, in turn, inciting criminal activity—or even mass destruction—against their targeted subjects? (Leventhal 1995).

Does it matter, for example, what the creator's intention was, or whether the game was designed as social commentary, postmodern art, or children's play? Are there limits to what can or should be represented as children's play, or bought and sold as playthings for mass audiences? How do we understand the nature of children's games—or high art masquerading as a children's game—when such representations can so easily become the provocative battleground over which a nation's, or a people's, popular hatreds can be visualized, actualized, toyed with, or enacted in the public sphere? These are some of the questions I intend to raise—and respond to—in the essay that follows. They are questions which are not so much about the ethics or psychology of play itself, as about the ethics of its commercially produced instruments of engagement.

So, let's take a closer look at the games in question. As previously mentioned, the games are from two distinct historical vantage points and two distinct genres of culture. The first is a commercially produced children's board game created in Nazi Germany during the Third Reich of the Holocaust era. While neither sanctioned nor created by official Nazi propagandists, the game draws explicitly on both their policies and their dehumanizing representations of Jews for its name, its board pieces, and its rules of play. The second is a work of fine art designed to look like a concentration-camp game, created over half a century after the historical event, and in an era of relative international peace. In what follows I will pay attention not only to the graphics and images associated with these two games but also to the essentially dehumanizing and sadistic implications articulated in the rules of play. Finally, I intend to situate the games in what Anthony Giddens calls their 'contextuality of action,' in order to fully understand their meaning of signification (1987: 99). 'Contexts,' Giddens argues, 'form "settings" of action, the qualities of which agents routinely draw upon in the course of orienting what they do and what they say to one another. Common awareness of these settings of action forms an anchoring element in the mutual knowledge whereby agents routinely make sense of what others say and do' (Giddens 1987: 99).

The first object of analysis is a modified cross and circle-style race game originally published in 1936 by a German firm named Gunther & Co. in Dresden, Germany.[2] It appears to have been first circulated widely in 1938, one month after *Kristallnacht* (the Night of Broken Glass), which kicked off a wave of anti-Jewish pogroms, violence, and destruction against the Jewish people, their synagogues, homes and places of business. Referred to as 'the most infamous board game in history,' the game was registered and marketed under the name *Juden Raus!* (Kick the Jews Out, or, more literally, Jews Out!), and was advertised as 'entertaining, instructive and solidly constructed.' As described by toy scholars Andrew Morris-Friedman and Ulrich Schadler,

> Juden Raus! is a race game that instills values of a totalitarian fascist regime. It has both the theme of racial hatred and employs racist images in the game design. The object of the game is to deprive the German Jews of their property and to make them leave the city" (Morris-Friedman and Schadler 2003: 2).

The game's equipment includes a pair of dice, a game board, and two types of game piece figurines. The 'men' figurines clearly represent the German police and can be identified by the coat and belt drawn on the pieces, as well as the buttons and boots which appear similar to uniforms worn by the German police at the time. The second type of game piece is a large cone-shaped hat with drawings of grotesque stereotypes of Jewish facial features, including long noses and menacing scowls, meant to represent Jews. These cone-shaped hats mirror those that Jews were compelled to wear during the Middle Ages (Fig. 10.1).

The board has 13 circles representing various Jewish-owned shops and businesses. Players take turns rolling the dice and moving their 'Jews' across the map from their places of business toward 'collection points' outside the city walls for deportation to Mandate Palestine. Written on the game board are the words, translated from German: 'If you manage to see off 6 Jews, you've won a clear victory!'—clearly foreshadowing the policy of genocide that was to follow (Morris-Friedman and Schadler 2003).

The second item I explore looks like a game, acts like a game, is labeled as a game, but is not a game. It is, rather, a postmodern work of art that was produced in 1996 by a Polish artist, Zbigniew Libera, over half a century after the end of WWII. His death-camp game has been shown in museums and galleries around the world, including New York's Jewish Museum in the controversial (2002) exhibition, 'Mirroring Evil' (Kleeblatt 2002). 'Lego Concentration Camp,' as the art piece is titled, was recently

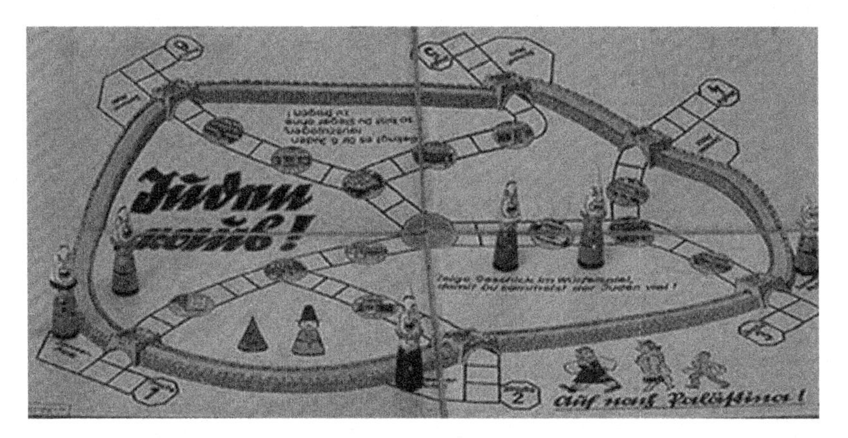

Fig. 10.1 *Juden Raus* board game. Owned by the Weiner Library, London

purchased by the Museum of Modern Art in Warsaw for a reported $71,800.00.[3]

The art work, of which there are three editions in existence today, comes in a set of seven individually boxed scenes, each featuring some visual aspect of concentration camp 'life,' including crematoria and barracks made from Lego bricks, as well as Lego figures dressed as 'guards' wielding sticks for beating 'inmates' (Fig. 10.2).

The Lego Corporation gave Libera the bricks for free without a clear vision of the project, and without knowing how he would use them. Although the Lego Corporation now insists that they did not endorse his art work, Libera's Concentration Camp art work has the words 'sponsored by Lego System' prominently stamped on each of the seven boxes.

Libera created the graphic design for each box by assembling Lego blocks into replicas of death-camp facilities and scenes, photographing them, and then using the photos to adorn the cardboard packages, each of which includes all of the Lego branding elements such as the company logo and safety warnings in multiple languages. The boxes graphically depict crematories, gallows, and doctor figures administering electric shocks to prisoners. One box cover illustrates a scene of random limbs piled outside

Fig. 10.2 Zbigniew Libera, Lego Concentration Camp, 1996. Collection of the Museum of Modern Art in Warsaw, photo by Bartosz Stawiarski

an Auschwitz-style barracks. In another, skeleton figures, taken from the popular Lego pirate series, haul bodies away to be incinerated.

Both of these games—the one created and designed to be a plaything for children, and the other created and designed to be a work of art for adults—were highly criticized upon their debut for 'trivializing' the annihilation of Jews at the hands of the Nazis—from the exact opposite perspectives. According to Morris-Friedman and Schadler, the *Juden Raus!* game was sharply criticized by the official SS newspaper, *Das Schwarze Korps*, for trivializing the nation's serious mandate to annihilate the Jews (2003: 9). Quoting from the newspaper itself, they note that the author's main objection is that the game manufacturer is trying to profit from the actual Nazi slogan, *Juden Raus!*, as a way of promoting the sale of the game.

> The political slogan 'Jews Out' is exploited here as a big seller for all toy shops and trivialized to an amusing pastime for little children!

> This invention is almost a punishable idea, perfectly suitable as grist to the mills of hate of the international Jewish journaille (gutter journalism), who would show around such a piece of mischief as proof for the childish efforts of the Nazistic Jew-haters with a diabolic smirk, if it would appear before her crooked nose.

Libera's Concentration Camp art piece was also criticized upon its initial presentations to the public, but by the voices of Holocaust survivors and mainstream Western media pundits who lambasted the piece for trivializing the inhuman horrors and trauma of that same annihilation.

So, what is going on here? Along with other forms of popular and visual culture, Nazi board games were an important element of Adolf Hitler's extensive propaganda campaign within Nazi Germany. Hitler's Propaganda Minister, Joseph Goebbels, understood that 'To be perceived, propaganda must evoke the interest of an audience and must be transmitted through an attention-getting communications medium' (Doob 1950). The Nazi Party used board games as one such medium to infiltrate the subtleties of everyday life and indoctrinate children with the National Socialist ideology. *Bomber uber England* (Bombers over England) and *Jagd auf Kohlenklau* (Hunting for the Coal Thief)[4] were two such popular games.[5]

To be clear, the use of games as popular culture propaganda during WWII was not limited to Germans, but was—and still is—a common tactic of all nations at war or facing perceived threats from those outside their borders. 'Bomb a Jap' and 'Trap the Jap' were two popular American games produced during the WWII era, as was the popular dartboard game from England with Hitler as the bullseye with a gaping black mouth which is marked for 100 points (Nelson 1990).

In each case, bold graphic images, insensitive language, and power-based rules of play are juxtaposed to encourage and teach children to participate in the 'game of war in two ways: by setting up the rules of the game to favor the allies over the enemies—a tactic common in both games and war—and by dehumanizing and creating hatred toward the supposed enemy by using derogatory or racist terms and images' (Findlay 2006).

This demonizing or de-humanizing process involved making individuals from the opposing nation, from a different ethnic group, or those who support the opposing viewpoint appear to be subhuman, worthless, or immoral, through suggestion or false accusations.

While the war games of Germany, Britain, and the USA could all be said to invoke stereotyped and racist images to depict their enemies, the *Juden Raus!* game depicted an internal victim—in this case over 6 million Jews—that was slated for annihilation through genocide. They were *not* leaders or even citizens of warring nations—but victims of an essentially racist eugenics movement implemented by the Nazi regime. In this sense, there might be more equivalence to the late-nineteenth-century US game, 'Ten Little Niggers' (1895), a racist variation of the Old Maid card game with exclusively African American characters, or 'Gunsmoke,' a cowboy-and-Indian game that incorporates racist themes in its rules of play by giving the Anglo cowboy side unfair advantage over the Native Americans (Nelson 1990: 20).

Toy scholar Pamela B. Nelson explores some of the most stereotyped toys produced during the last 150 years of America's history, and the direct correlation between their mass production and the rising anti-ethnic or race-based sentiment of the nation. The 'Reclining Chinaman' is one such example of a mass-produced mechanical toy from the 1880s that, she notes, 'reinforced the image of the Chinese as crafty tricksters who cheated American working men out of jobs by accepting lower wages and an inferior standard of living' (Nelson 1990: 11). While America did not find itself at war with China during this time, the country's widespread anti-Chinese sentiment did lead to the passage of immigration laws virtually banning the immigration of Chinese laborers to the United States—the very same year that the mechanical toy was produced and distributed as a popular plaything for children.

The *Juden Raus!* game similarly presaged a very real practice known as *Judenjagd* (hunt the Jew), which involved searches for Jews who had survived ghetto liquidations and deportations to death camps in Poland in 1942 and had attempted to hide 'on the Aryan side.' The ultimate goal of such 'hunts' was the brutal evacuation and extinction of the hunted and discovered Jews in the Nazi death camps.[6]

As we have seen, the *Juden Raus!* game was literally saturated with symbols and icons that directly relate to centuries-old cultural narratives of Jews as crooked businessmen, wily, unruly citizens, and evil and devilish-looking characters. The rules of play also mirror longstanding political and ideological practices of isolating, exiling, or extinguishing Jews from as far back as the Christian Inquisition. So effectively did the rules mimic actual policy, as we have seen, that Nazi officials condemned the game as confusing the public as to the very serious nature of their own *Juden Raus!* initiative. While each of the visual aspects of the game can thus be read and interpreted in the immediate context of Nazi Germany's actions, their meaning must also, ultimately, be understood within a centuries-old, shared public discourse of dehumanization and demonization of Europe's Jewish population by the dominant Christian culture. They form part of what cultural theorist Antonio Gramsci would call the largely unconscious and uncritical—common-sense—way of perceiving the world, and the seemingly 'natural' order of people in that world, that is so widespread in any given historical epoch (1971: 322–326).

Now I would like to move forward in time to the other death-camp 'game' in question, the Lego Concentration Camp set that was produced by a Polish artist in 1996. The artist, born in 1959 into the immediate postwar generation, had been raised on a steady stream of Holocaust-related popular culture presentations, including movies, books, graphic novels, lectures, plays, and museum exhibits.

Most of these media pieces fall into the category that cultural scholars have termed 'trauma media,' a genre articulated by philosopher Berel Lang who argues in his book *Act and Idea of the Nazi Genocide* that there are only certain appropriate and ethically responsible ways of representing the Shoah (in Leventhal 1995). Unlike other art that can claim autonomy or self-reflexivity, Holocaust art, Lang and others suggest, should respectfully promote Holocaust education and remembrance (van Alphen: 71). In consequence, there are unwritten rules of representation that it should follow, including a rigid faithfulness to facts and historical conditions, a commitment to a serious and almost 'sacred' approach that above all honors the victims and the dead, and a portrayal of the Holocaust as a unique event, separate from all other acts of genocidal history (van Alphen: 73).

Lang and others have suggested that the Holocaust has also tended to require what he calls an 'elevated' genre of representation—that it is the stuff of 'high' art and should not be 'desecrated' by allowing low genres to communicate the destruction of the European Jews. 'Toy' art, they argue, is not 'serious' enough in terms of historical reconstruction.

Indeed, Art Spiegelman's Pulitzer Prize-winning graphic novel *Maus* (1986) was initially criticized for desecrating, or trivializing, the Holocaust, by presenting such a serious topic in such a conventionally juvenile or lighthearted genre. But unlike the Lego Concentration Camp set created by Libera, there is no ambiguity of who the protagonist/hero of *Maus* is—the author's Jewish father; not the Nazi perpetrators (Leventhal 1995).

Scholars have noted that the Lego Concentration Camp boxed set essentially ruptures the conventions of trauma representation both through genre and point of view: a seemingly purposeful trivialization of the horror of the camps, and a blurring of roles for the player/viewer charged with vicariously constructing the camp and carrying out its unspeakable acts. With Libera's toy Lego set, the viewer is encouraged to envision him-or herself in a situation comparable to the 'real' situation, and in the role of perpetrator, rather than victim. Identification, remarks scholar Ernst van Alphen, 'replaces mastery' and the victimizer replaces the victim as the object of that identification (van Alphen: 75). Libera's Lego Concentration Camp set, he maintains, 'stimulates visitors of the museum or the gallery to envision the possibility of building their own concentration camp.'

Scholars of Libera's Lego set ask, pointedly, 'What happens when we play with traumatic histories?' Some, like van Alphen, suggest that such play may actually serve a crucial pedagogical role in Holocaust remembrance and education, especially for the third and fourth generations of Holocaust bystanders and survivors who have no direct relationship with the event. For such students of the Holocaust, Libera's fundamentally affective mode of pedagogy may actually be the very thing that allows current generations to identify sufficiently with the perpetrators of evil to recognize 'the ease with which one can slide into a measure of complicity.' To play-act the Holocaust in this way, under the strict direction of an artistic 'director,' they suggest, allows these media-saturated post-Holocaust generations to emotionally experience a trauma they have not directly lived and thereby work through essentially sadistic, horrifying aspects of history and human nature.

The artist himself corroborates this point of view, claiming that neither he nor this work is anti-semitic, or even directly related to the Holocaust of Nazi Germany. He insists that those who disapproved of his Lego piece were confusing art with life—mistakenly thinking that 'this was not an art work but a real toy.' Rather, he says, the Lego Concentration Camp piece should be read 'metaphorically,' as a productive provocation of child rearing and cultural norms that regulate the marketplace and shape public consent. 'This is significant,' Libera opines, 'because currently wars are fought not only with weapons but also with products and culture' (in Marcoci 1998).

> ...it is obvious that when we play, we do not want to identify with the victim. So then let's identify with the executioners! Emotions are important and they play a more important role than rational thinking in this process. This is what the Greeks called a Dionysian rite. They used their emotions more than their rationality. We do not really know what to do with emotions. Plus, today you are supposed to create only and exclusively the 'right' emotions. How should you do that? I do not know. I am not in the Shoah business, but I am somebody who reveals that a Shoah business exists.[7]

Critics of Libera's Holocaust toy art counter that such cultural pieces—whether actual games or postmodern artistic spoofs on such games—are anything but benign. As the founding chairman of the International Network of Children of Holocaust Survivors remarks, 'What can a Lego concentration camp mean, except that killing is child's play?'[8]

Indeed, it would be only a matter of time after the appearance of Libera's Lego Concentration Camp set that actual Holocaust-themed games would find their way into the global marketplace, and Libera's insistence on the distinction between 'toy art' and 'real toys' would be painfully tested. And the intellectual question he sets forth in his art work about the nature of human behavior (Could I really play this game? Could I participate in something as terrible as the Holocaust?) would be answered by the thousands who play the games themselves, and the hundreds who have perpetrated actual acts of violence mirroring the game's commands.[9]

Unlike Libera's work, these games *are* actually playable—creating a real-life context in which someone in post-Holocaust culture performs or play-acts Holocaust events leading up to and within the setting of a camp.

One such board game, called simply 'Train,' was designed in 2009 by Savannah College of Art and Design professor, Brenda Brathwaite, as an educational tool for teens and young adults. The board game involves the task of loading little yellow people onto trains for some undisclosed final destination, only late in the game revealed as Auschwitz. At that time, players have an option to try to 'save' the people from the horror that, based on the historical reality, they now recognize awaits them.[10]

The key emotion that Brathwaite said she wanted the player to feel was 'complicity.' 'People blindly follow rules,' she said. 'Will they blindly follow rules that come out of a Nazi typewriter?'[11] What is interesting about Brathwaite's game is that, for her, it is more of a social experiment than a game—and she insists on controlling each and every incidence of its playing. Indeed, she describes one incident in which one set of players—some kids who had had too much to drink—became 'disrespectful,' and she stopped them in mid-play, forbidding them to continue with the game/experiment.

Brathwaite maintains that games such as hers do not trivialize the subject matter or show any disrespect to the actual historical events they depict as child's play. Rather, she opines, 'I think games are a good medium for approaching any subject, particularly difficult ones, because by their very nature, they are abstract, invite interaction and allow us to confront and question things … particularly rules that we may blindly follow' (in Brophy-Warren 2009).

Yet what happens when the contexts for playing such games are not so carefully controlled, and their creator's intent, once the game 'goes viral,' becomes beside the point? One such game—an online simulation game called 'Auschwitz Concentration Camp'—surfaced as recently as June 2016. According to the Italian newspaper La Repubblica,[12] the game was developed as a smartphone app that allows users to 'live like a real Jew' in a concentration camp, and features ovens used to burn bodies of Jews murdered in the camp. The app, which had a 'Teen' rating for 'violence and blood,' was marketed as 'a macabre game that replicated the Holocaust in a digital version.' The homepage features a Star of David alongside train tracks with the words 'Auschwitz Concentration Camp' superimposed above.

The free downloadable game was created by a vocational school specializing in the construction of computer games in Zaragoza, Spain. Intended, they say, as 'just a parody,' the producers claimed that their goal was never designed to perpetuate hate or incite violence, but rather to see how easy it was to get such a clearly racist game on platforms such as Google. The game was uploaded by the makers of the game and remained on the Google site for just over a week (until Google removed it amid wide public condemnation). Even in that short time span, the game acquired several thousand fans, and had an online review rating of 3.1 stars out of 5.

Despite what the producers may or may not have intended, the game-users' responses were horrifically disturbing. The World Jewish Congress noted that 'one user expressed his enjoyment at playing the game and said that the only problem was "that every 20 min I find the full oven and I have to come to remove ash."'

While Brathwaite and others continue to insist, as did Libera before them, that such games—or games as art or social experiment—do not trivialize or capitalize on the horrors of the Holocaust, I would argue that they are, in fact, part of a global discursive formation—perpetuated on the level of ideas, images and practices—which continually reinforces and perpetuates the idea of the 'Jew' as a reviled and demonized 'Other,' ripe for extinction or exile. Such discourse, critical theorists understand, is not something that is consciously perpetuated, performed, or enacted, but rather something operating much more insidiously on the level of 'common sense': what Gramsci refers to as 'the largely unconscious and uncritical way of perceiving the world that is widespread in any given historical epoch.'

The danger of such discursive formations—including death-camp games—becomes painfully apparent when they begin to foreshadow real life forms of violence that they so 'playfully' depict. This was certainly the case during the Third Reich of Nazi Germany, and has, unfortunately, become increasingly apparent throughout Europe and the Middle East today, where headlines portray a startling rise of anti-semitic speech and acts of violence in every country throughout Europe. It is in this context that we must acknowledge the ease with which such ubiquitous discourses of child's play find real-life parallels in acts of violence against the same populations and properties where such persecution was historically incited.

What might have begun in Libera's time as a more or less intellectual discussion about the power of art, masquerading as child's play, to teach lessons about the nature of evil in an age of mass production, has escalated in the intervening two decades into a full-blown crisis of oppression against Jews incited and perpetrated on a global stage. 'After a lull of more than 70 years,' writes critical commentator Menachem Rosensaft, 'anti-Semites in a succession of European countries now unabashedly shout and brandish signs proclaiming "Death to the Jews" and "Gas the Jews." And even the now familiar "*Juden Raus!*" has returned as a common epithet spray painted or graffitied, often together with a swastika, along the walls of public places of business, bridges, underpasses, cemeteries and building facades, presumably to leave no doubt as to the vandal's mindset or intent' (Rosensaft 2016).

Across Europe, synagogues, restaurants, and Jewish places of business are again under attack, and physical violence perpetrated against Jews is steeply on the rise. 'These are the worst times since the Nazi era,' remarked the president of Germany's Central Council of Jews in the summer of 2014. 'On the streets, you hear things like "the Jews should be gassed," "the Jews should be burned"—we haven't had that in Germany for decades. Anyone saying those slogans isn't criticizing Israeli politics, it's just pure hatred against Jews: nothing else. And it's not just a German phenomenon. It's an outbreak of hatred against Jews so intense that it's very clear indeed.'[13]

It is against this backdrop that the recent proliferation of anti-semitic images, cartoons, games, and toys must be analyzed, and Libera's prescient Lego Concentration Camp set must be historically situated. As Rosensaft correctly notes, 'Attacks on Jews, whether by neo-Nazis or pseudo-Jihadists, do not occur out of nowhere. They are almost always the result of long-term indoctrination by hate speech' (Rosensaft 2016).

And the games which embody this hate speech in both their representation and rules of play? Make no mistake, he insists, 'Playing with hateful acts is just one step away from committing the act itself.'

> If 'incitement [to genocide] is a step toward genocide,' as Susan Benesch, director of the Dangerous Speech Project, has observed, then I would argue that inciting discrimination and violence against a targeted group constitutes an equally significant, perhaps even inevitable, step toward incitement to genocide.

Cultural studies scholars, including Antonio Gramsci, Michel Foucault, Raymond Williams, and Stuart Hall, and beyond, offer insightful analyses about the specific ways in which popular culture items not only reflect a society's dominant values and positions about a hated or feared 'other,' but work subliminally to both create and maintain them in the national public imaginary. One of the most useful aspects of Gramsci's theory of hegemonic cultural production, in fact, is that it utterly refutes the suggestion that the relationship between popular culture and a society's beliefs and values is merely reflective. Indeed, for Gramsci, the relationship between culture and a society's politics and values is a process, never stable, which must be constantly negotiated, challenged, fought over, stated, and protested—often in the realm of movies, magazines, posters, and popular culture items like toys and games. Games—or high art parading as games—must be seen in this construction not only as a reflection of a country's dominant acts and ideologies, for example, but as active components in the constitution, reinforcement and entrenchment of those ideologies.

While the late-twentieth- and twenty-first-century creators of Holocaust toys and toy art may insist that their creations are mere parodic commentary—or cautious education—on the nature of evil in our lives, repeated events of history teach us that, in fact, they are playing with dangerous fire. Even if a postmodern parody such as Libera's Lego Concentration Camp set could have been sufficiently divorced by time from the need to approach the Holocaust with sacredness and solemnity, that time has passed. With resurgent anti-semitic views and violence across Europe, and with the instantaneous and anonymous ability for the Internet to become a home for such anti-semitic views, the context has changed again. In the 1930s, German parents buying the *Juden Raus!* game could have believed that their children's consumption of hate-motivated games was benign. Today, the comingling of play and hate that comes into children's Internet-linked devices should raise alarm, not because play is innocuous, but because play and playthings are all too powerful.

Notes

1. Wikipedia: Wartime Propaganda. https://en.wikipedia.org/wiki/Propaganda.
2. *"Jews Out Board Game" History in an Hour. 6 July 2012.*
3. As reported in the Culture Monster section of the *Los Angeles Times* by Dawn Marie Murphy on January 3, 2012.
4. *"Jagd auf Kohlenklau (Hunt the Coal Thief)". The British Museum.*
5. *Dolan, Andy (2007-08-20). "German children played with 'Bombers over England' boardgames during WWII". The Daily Mail.*
6. See Jan Grabowski's excellent microhistory, 'Hunt for the Jews: Betrayal and Murder in German-Occupied Poland,' for a full account of the *Judenjagd* practice in one rural county in southeastern Poland where the majority of Jews in hiding perished as a consequence of betrayal by their Polish neighbors.

7. As quoted in Hedvig Turai, "The Artist Does Not Own His Interpretations: Hedvig Turai In Conversation with Zbigniew Libera. ArtMargins online. May 25, 2006.
8. Menachem Rosensaft, as quoted in Alan Cooperman's "Art or Insult: A Dialogue Shaped by the Holocaust." The Washington Post, February 24, 2002.
9. So common have online, virtual Holocaust-themed games become over the past decade, that the genre itself would be parodied in an online amateur comedy site called 'uncyclopedia' in 2006. Under the title 'Holocaust Tycoon,' the parodied game is described as a simulation computer game, released for Windows XP and Wii platforms in 2006. It was developed by Ka-tzetnik Enterprises in Berlin, and later released throughout Europe by WN Software. In Holocaust Tycoon, the player must successfully manage a camp, without going bankrupt, whilst avoiding the attention of invading Allied Forces.
10. Brophy-Warren. "The Board Game No One Wants to Play More Than Once." *Wall Street Journal*, June 24, 2009.
11. As quoted in Dean Takahashi, Brenda Braithwaite Romero's "Train" board game will make you ponder. *VentureBeat.* May 11, 2013.
12. www.repubblica.it/.../app_antisemita_auschwitz_spagna-142233096.
13. Guardian 2014.

References

Bacon, J. (2013). Review of war games exhibition at Victoria and Albert Museum of Childhood. Retrieved from http://www.forbiddenplanet.co.uk/blog/2013/exhibitions-war-games/.

Brophy-Warren, J. (2009). The board game no one wants to play more than once. *Wall Street Journal.* Retrieved from http://www.blogs.wsj.com.

Cooperman, A. (2002). Art or insult: A dialogue shaped by the Holocaust. *The Washington Post.* Retrieved from https://www.washingtonpost.com.

Custodero, A. (2016). App antisemita simula la vita degli ebrei ad Auschwitz. *La Repubblica.* http://www.repubblica.it/argomenti/auschwitz.

Dolan, A. (2007). German children played with 'Bombers over England' boardgames during WWII. *The Daily Mail.* Retrieved from http://www.dailymail.co.uk.

Doob, Leonard W. (1950). Goebbels' principles of propaganda. *The Public Opinion Quarterly, 14*(3), 419–442.

Feinstein, S. (Ed.). (2005). *Absence/Presence: Critical essays on the artistic memory of the Holocaust.* Syracuse, NY: Syracuse University Press.

Foucault, Michel. (1980). *Power/Knowledge.* Colin Gordon.

Findlay, James A. (2006). *The game of war: books, toys, and propaganda from the Mitchell Wolfson, Jr. Study Centre.* Fort Lauderdale, FL: Bienes Center for the Literary Arts. Retrieved from http://www.broward.org/library/bienes/documents/gameofwar.pdf.

Giddens, Anthony. (1987). *Social theory and modern sociology*. Stanford, CA: Stanford University Press.

Gordon, C. (Ed.). (1980). *Power/Knowledge: Selected interviews and other writings, 1972–1977 of Michel Foucault*. New York: Pantheon Books.

Grabowski, J. (2013). *Hunt for the Jews: Betrayal and murder in German-Occupied Poland*. Bloomington, IN: Indiana University Press.

Hall, Stuart, Evans, J., & Nixon, S. (Eds.). (1997). *Representation: Cultural representation and signifying practices*. New York, NY: Sage Publications.

Jowett, Gareth, & O'Donnell, Victoria. (2006). *Propaganda and Persuasion*. New York, NY: Sage Publications.

Kleeblatt, N. (2002). *Mirroring evil: Nazi imagery/recent art*. New Brunswick, NJ: Rutgers University Press.

Leventhal, Robert S. (1995). Art Spiegelman's MAUS: Working through the trauma of the Holocaust. Retrieved from http://www2.iath.virginia.edu/holocaust/spiegelman.html.

Marcoci, Roxanna. (1998). The antinomies of censorship: The case of Zbigniew Libera. *Index* 3–4.98, no. 23: 58–59.

Morris-Friedman, Andrew & Ulrich Schadler. (2003). 'Juden Raus!' (Jews Out!): History's most infamous board game. *Board Game Studies* 6. Retrieved from http://www.academia.edu/2149566.

Nelson, Pamela B. (1990). Toys as history: Ethnic images and cultural change. In Pamela B. Nelson (Ed.), *Ethnic images in toys and games: An exhibition at the Balch Institute for Ethnic Studies, April 17-October 13, 1990* (pp. 10–18). The Balch Institute for Ethnic Studies: Philadelphia, PA.

Nelson, Richard Allen. (1996). *A Chronology and Glossary of Propaganda in the United States*. Westport, CN: Greenwood Press.

Rosensaft, M. (2016). On the origins of anti-semitism, which rages today. *The Scroll*. Retrieved from http://www.tabletmag.com/scroll.

Spiegelman, A. (1986). *Maus: A survivor's tale*. New York: Pantheon.

Takahashi, D. (2013). Brenda Romero's train board game will make you ponder. *VentureBeat*. Retrieved from http://www.venturebeat.com.

Turai, H. (2006). The artist does not own his interpretations: Hedvig Turai in conversation with Zbigniew Libera. *ArtMargins*. Retrieved from www.artmargins.com.

Van Alphen, E. (2002). Playing the Holocaust. In Kleeblatt, N. (ed.). *Mirroring Evil: Nazi imagery/recent art*. 65–84.

Wartime Propaganda. Retrieved from https://en.wikipedia.org/wiki/Propaganda.

Whitehill, Bruce. (1999). American games: A historical perspective. *Board Games Studies, 2*, 116–141.

Author Biography

Suzanne (Suzy) Seriff (Ph.D.) Dr. Seriff is a Senior Lecturer in Anthropology at the University of Texas at Austin, and a public folklorist working at the intersection of folk arts, community activism, and social justice. She has been a member of ITRA since 1999 and a board member since 2008. She is the co-editor of ITRA's biannual newsletter.

CHAPTER 11

Working Class Children's Toys in Times of War and Famine. Play, Work and the Agency of Children in Piraeus Neighborhoods During the German Occupation of Greece

Cleo Gougoulis

Introduction

Children's experience of war has come under the scrutiny of a broad range of childhood scholars from various disciplines, both in the context of a discussion of war and conflict trauma and a growing interest in the diversified meanings of childhood as a social and cultural category since the publication of the pioneering work by James and Prout (1990). Prevailing notions of Western childhood as a period in the life course characterized by social dependency, asexuality, and the obligation to be happy (Ennew 1986) clash sharply both with the destructive nature and the pragmatics of war, especially in cases where children are recruited as child soldiers.

C. Gougoulis (✉)
University of Patras, Patras, Greece
e-mail: cleogougoulis@yahoo.gr

L. Magalhães and J. Goldstein (eds.), *Toys and Communication*,
https://doi.org/10.1057/978-1-137-59136-4_11

While literature on child soldiers in particular raises questions on the role of children as victims and perpetrators of violence (e.g. Honwana 2005; West 2000), recent works also explore the experience of war and the methods employed by adults and children for coping with war and its trauma. Kozlovsky (2008) has explored the convergence of trauma theory with an educational interest in play in the implementation of social policies, such as the introduction of adventure playgrounds in postwar London, aiming to alleviate the detrimental effects of warfare experience on children. Play and toys form part of these methods of coming to terms with distress and the horror of war in the light of psychoanalytic readings of play as cathartic and therapeutic (e.g. Moshenka 2008), while the connection of play to celebration and resistance against the unthinkable Nazi atrocities of the Final Solution permeates the compelling work of George Eisen (1988) on Jewish children's play in the Holocaust. The vitality of children playing against the odds in the face of death is astonishing. The same determination to live in the face of the famine outbreak during the German Occupation of Greece has been recorded by autobiographical material and more recently in oral history projects[1] run by contemporary historians of war and trauma, thus raising the question of the importance of play and playthings as a strategy for claiming life. Play was a vital part of children's daily routines next to chore performance and other informal forms of labor, during the famine sweeping many urban centers in Greece, most notably Athens and Piraeus. Children played throughout this period by making their own play materials 'out of nothing' and yet children's play remains under-represented in published scholarly work on childhood during the war.

This chapter will deal with the ingenuity and inventiveness exhibited by working-class children during World War II in an urban milieu. The material is based on interviews and toy-making sessions with the elderly in two Day Care Centers (KAPH) situated in two working-class neighborhoods in the municipality of Nikaia, Piraeus, Greece (see Fig. 11.5), an area renowned for its massive participation in the anti-Nazi resistance and inhabited during the war by Greek and Armenian families of refugees from Asia Minor who settled in the area after the military defeat of the Greek Army in Turkey in 1922 (see Hirschon 1989: 35–48). The toy-making sessions were organized in the context of an exhibit under the title 'Toys from the Attic' that was mounted in January 2014 in a local primary school as part of ethnographic research I have been carrying out in the area. The exhibit was the outcome of an outreach program initiated by the author as a method of collecting stories about people's relationship to their toys and the context of their use. The project encouraged intergenerational exchange between the local elderly, parents, teachers, and students, both during the preparation and in the aftermath of the exhibit when toy-making workshops were run by the elderly members of the local KAPH.

The stories about toys during WW II were collected in this context before and after the exhibition by interviewing 22 KAPH members (13 men and 9 women) born between 1927 and 1940. Adult reminiscences have been critically examined and compared to other written and oral sources in the light of a growing literature on oral history and the problematic nature of the relationship between individual memory, collective memory and history (e.g. Halbwachs 1992; Abrams 2010; Bergen and Niven 2014) and discussions on the difficulties of reconstructing childhood stories from adult memories (e.g. West and Petrick 1992).

Toys During the Nazi Occupation of Greece

WWII began in Greece on October 28, 1940 with the invasion of its north-western border by the Italian Fascist Army within 12 h after the Greek government rejected Mussolini's ultimatum to surrender to the Axis. A massive mobilization followed within these 12 h, as Greeks flocked to the colors, and in fact the Greek army did manage initially to push the Italian invaders back and to delay the German Nazi plans in the Balkans. This success was short-lived, lasting for some months until the Nazi troops finally marched into Athens in April 1941 and raised the swastika over the Acropolis (Mazower 2001: 1–5). Nevertheless, the victory over Mussolini's army has been vividly preserved in national memory since 1941 through the celebration of 'OXI' day ('NO' day'), the day that the Greek people said 'No' to Mussolini's ultimatum, first in the form of anti-fascist demonstrations and strikes in Nazi-occupied Athens, and after the liberation in the form of national parades on October 28.

The historian Mark Mazower (2001: 1), drawing on the memoirs of the Greek novelist George Theotokas (n.d.: 260), gives a lively picture of the public enthusiasm prevailing in the streets of Athens during the first months of the war, when 'excited crowds shouted insults against the Axis, while wounded soldiers, convalescing in hospitals sang the national anthem and demanded to be sent back north to fight.'

One of the paradoxes of war is that, on the one hand, it has devastating consequences for the affected populations in terms of the death toll and the trauma experienced under acute conditions of adversity while, on the other, it mobilizes people to resist destruction and, among other things, invent survival strategies. Toys have been known to be part of these survival strategies, often constituting a form of political propaganda aiming to boost the nation's morale. Greek toys produced during the war were no exception to this rule. While most of the toy-manufacturing units closed

Fig. 11.1 Wooden Mussolini puppet peddled in the streets of Athens 1940–1941. Photo by Voula Papaioannou © Benaki Museum Photographic Archive. Published with permission

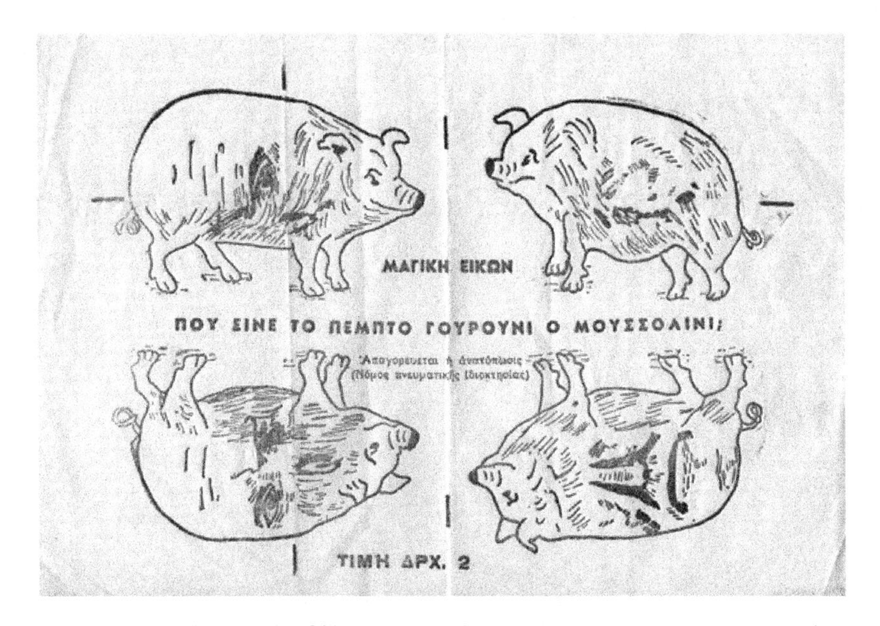

Fig. 11.2 "Where is the fifth pig, Mussolini?" Paper toy caricaturing the Italian fascist leader Mussolini in a dehumanizing way. Toys and Childhood Museum, Benaki Museum, index nr. ΤΠΠ 5035 1–2. © Benaki Museum Copyrights, Athens 2016. Published with permission

down during the Nazi Occupation, some toys were still made by craftsmen and peddlers as part of a diversified survival strategy (Argyriadi 2008: 167).

Greek toys in the period of the outbreak of WWII both reflected and contributed to the dissemination of the idea of the people's resistance. Puppets depicting the Italian dictator Mussolini shown in Fig. 11.1 were peddled in the streets of Athens and paper cuts portraying the fascist leader in the form of a pig (see Figs. 11.2 and 11.3) aimed at degrading the enemy (Argyriadi 2008: 133).[2]

Famine and Family Strategies of Survival

One of the dire consequences of the German Occupation of Greece was the outbreak of an unprecedented famine in Athens and many cities of the province during the winter of 1941–1942 and in Piraeus 1941–1943. The famine was the result of several causes, the most prominent being the Nazis'

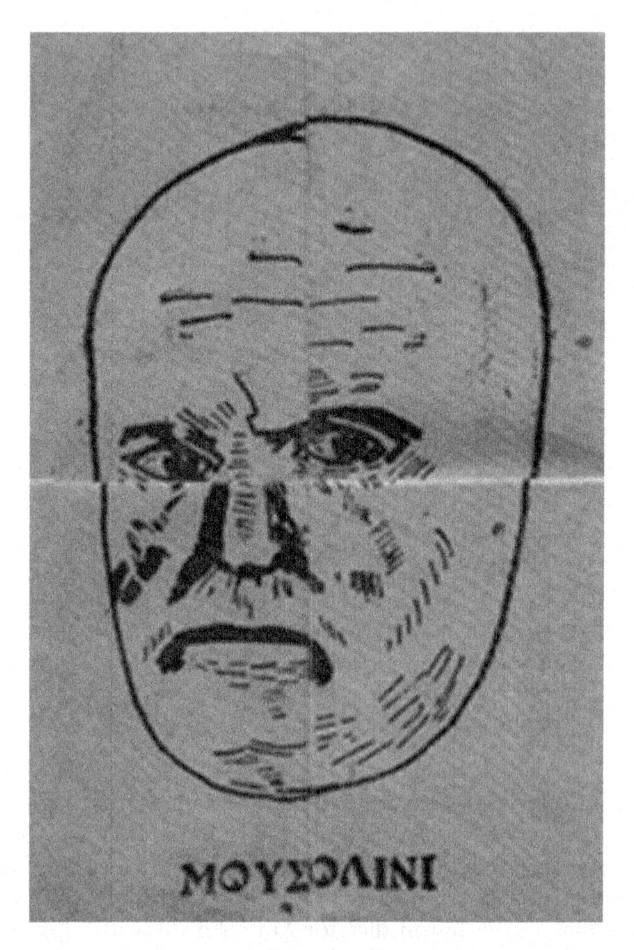

Fig. 11.3 "Where is the fifth pig, Mussolini?" Paper toy caricaturing the Italian fascist leader Mussolini in a dehumanizing way. The hidden portrait of Mussolini is revealed by folding the paper along the designated lines. Toys and Childhood Museum, Benaki Museum, index nr. ΤΠΠ 5035 1–2. © Benaki Museum Copyrights, Athens 2016. Published with permission

refusal to adhere to international treaties concerning the survival of conquered populations. Most food supplies and the greatest part of agricultural production were confiscated for the needs of the German Army. Another reason for food shortage was the inability of the puppet Greek government during the Occupation to meet the acute demand for food. Furthermore, Greece mainly relied on imports for food, so food shortage was partly due to the naval blockade imposed by the British to occupied countries as a form of pressure against the Nazis. It took about a year of combined efforts and tens of thousands of deaths in Athens and Piraeus alone to convince the British Government to lift the blockade for ships of the Red Cross carrying food supplies (Hadjiiossif 2007: 197; Loukos 2007: 220, 231).

The experience of famine dominates all childhood memories of my research participants. When prompted by the researcher on the existence of commercial toys during the war, most participants stated emphatically that people were dying in the streets, there was nothing to eat, olive oil and flour were scarce, let alone toys ...

Famine was so overwhelming that some families refused to report their dead, in order to continue to use their ration cards for the Red Cross. George Z. still remembers with horror and regret that, when his father's brother died, the family lowered his body over the cemetery wall, as a proper official burial would mean loss of one food ration for the famished family.

How did families survive the extreme food shortage?

Some parents engaged in retailing. They travelled to adjacent provincial towns which still had some supplies and exchanged clothes and linen for raisins and corn. Some had to trade a sewing machine for a loaf of bread. Trading on the black market was widespread. Parents working in German barracks or Nazi-occupied public services and enterprises as cooks kept potato peelings for their children, while others kept portions of food from lunchbreaks at factories where food was distributed to workers.

Children also contributed to the daily struggle of finding food by stealing, looting, begging, and running errands.

George Z. and his brothers used to loot a fruitmonger's storerooms filled with onions. The biting taste of onion is still etched in his memory. Others, the renowned *saltadoroi*, leapt over passing German trucks carrying bread and other supplies. Some childhood memories focus on stealing fruit from gardens at the beginning of the war.

Play was closely linked to food search, especially in memories of boyhood throughout the war, as food shortage and poverty were still felt in the area throughout the war and the ensuing Civil War 1946–1949 (see Desypris 2016: 149–163) as in other parts of Greece (Hionidou 2011).

Grigoris D. and Vangelis S. remember how they used to dig holes under the carob trees to find carob seeds, which were a real treat, as their mothers ground them and used carob seed powder for cooking pies. Hunger was a haunting feeling throughout the period 1941–1949 in the poorer districts, so children often resorted to chore making to gain a slice of corn bread.

Children's Portrayals and the Experience of Childhood During the War

One of the strategies for mobilizing international food relief programs was the systematic portrayal of famished children in portfolios by Greek professional photographers, such as Voula Papaioannou (see Fig. 11.4), compiled on behalf of the International Red Cross and the Near East Foundation in an attempt to exert pressure on international public opinion and convince the Allies to end the blockade.

Fig. 11.4 Portrayal of starving child during the famine in Athens Greece, 1941–1942. Photo by Voula Papaioannou © Benaki Museum Photographic Archive. Published with permission

Children in these portfolios appeared as the main victims of the war and one of the biggest campaigns for food relief launched by the Near East Foundation was entitled 'Save our babies' (Konstantinou 2007: 19–21). While children did suffer, and famine features prominently in all childhood recollections of my informants, the prevailing concept of childhood as a period deserving protection and care permeating all social strata by the time of the war did in fact prevent a steep increase in child mortality due to starvation. It was male adults, the jobless and the elderly who were most affected according to recent historical studies of deaths in Athens and Piraeus during the famine (Bournova 2005; Loukos 2007: 220; Desypris 2016). Parental efforts were not as effective in preventing the rise of infant mortality—infant mortality rates in the area tripled during 1941–1944—but the overall proportion of child deaths of famine amounted to 11.9% of the male and 10.5% of the female population's deaths from famine in the area between 1941 and 1944 (Desypris 2016: 154–155).

Social historians and anthropologists since the pioneering work of Philippe Ariès (1973) have argued that childhood is a socially, culturally, and historically dependent category, rather than a universal biological stage in the life cycle (Cunningham 1995; James et al. 1998; James and Prout 1990).

Children seemed to grow up sooner during the years of the Occupation. In many of the life histories collected for National Greek Television and studied by the historian Odette Varon-Vasar, informants claimed that their childhood ended with the end of primary school at 10 or 12 years of age (Varon-Vasar 2009: 74).

Thousands of teenagers, some as young as 14 years of age, actively participated in the resistance by enlisting in EPON, the youth wing of the National Liberation Front (an organization affiliated to the Greek Communist Party and the biggest resistance organization). Younger children, the *Aetopoula* (Little eagles), were also recruited as an ideological preparation for entering EPON and supported the resistance by forming an information network, but there were no child soldiers in the Greek Resistance. Children younger than 14 have not been recorded as being involved in armed warfare during the Nazi Occupation (Lyberatos 2007: 28; Vervenioti 2005).[3]

While none of my informants was organized during their childhood in a resistance group, most of them had family members involved, some of whom were arrested, imprisoned, or executed by the Nazis. Childhood among my informants was experienced as a complex period of the life cycle with obligations to the immediate family but also as a time of autonomy and play. Dependency on parents for food and subsistence seemed to last until a fluctuating age range between 9 and 12 years. Boys and girls from the age of 9 or 10 years onwards, depending on the duration of their attendance and the age of graduation from primary school, worked for their livelihood, helping parents, while younger children resorted to chore making to gain some extra food, money or play materials. Jobs for boys usually consisted in working as street hawkers, or as apprentices among local craftsmen (e.g. shoemakers, stonemasons, metalworkers), while girls worked as apprentices for seamstresses and mechanical loom workshop owners. Both genders worked at local factories and participated in strikes and other forms of struggle for better payment or food rationing organized by 'People's committees' and factory committees during the war (see Mazower 2001: 108–110).

Children did find time to play—especially as schools were closed for long periods during the war, due to frequent air bombardment in the area. Street play continued for both genders well into adolescence and well beyond the age of marriage among girls, who married as young as 16.

Play and Toys During the Occupation

The fact that toy production came to a standstill and many manufacturing units closed as a result of the war, and the long absence of most of the economically active male population during the Greek-Italian War (Argyriadi 2008), did not stop children's toy play during the war. Commercial toys were scarce even before the war in the working-class neighborhoods, where some children had never seen commercial dolls or toys enjoyed by the middle-class clientèle of the Athenian department stores before the war. And yet some inexpensive playthings, such as marbles, balls, spinning tops, and scraps that had been available in local corner shops before the outbreak of famine, were still present in some children's homes and made their way into the streets and *alanes* (empty plots of land, or urban wastelands), the play spaces of Greek urban working-class children during the greater part of the twentieth century.

Minas A. remembers that some children had commercial toys, e.g., the son of his boss, a goldsmith. George Z. and Mihalis P. remember that some godparents managed to find rubber balls for their godchildren for New Year's Eve even during the war.

For the vast majority of poverty-stricken children of the neighborhood, however, commercial toys were out of the question. If godparents, the traditional gift bearers for children at New Year's Eve and Easter, could bring some kind of present, it was more likely to be shoes than toys. Shoes —if available—were treasured and kept for special occasions. One woman, Evdoxia, recalled making her own shoes out of rags, but most of the elderly informants' descriptions of children's attire focused on walking and playing barefoot, with patched clothes and rags, in the streets and the *alanes*.

Memories of boyhood in particular focus on blistered feet and injured knees. Despite the harsh reality of the war, children did manage to play, and life did go on for those who managed to survive the famine through mobilizing their networks and all members of the family to work for their living.

Most of the toys mentioned in interviews during this era were made by children themselves, parents and relatives. The overriding majority of these toys were not mentioned as cherished possessions but as play instruments in games, some of them being short-lived constructions made anew for each play session. Toys, as Brian Sutton-Smith (1986: 26) has argued, have been historically less important to children than playing with each other. While this seems to be the case for the working-class children of Piraeus neighborhoods, it does not necessarily mean that children did not own toys. Boys proudly carried their cloth bags filled with clay marbles and a few precious glass marbles used as shooters, but toy ownership was not a means for gaining status. As toys were integrated either in pretend play scenarios or in traditional rule-governed street games, my descriptions of toys will be structured according to their use during play.

Gender Relationships and Play Groups

Girls and boys played at some street games together in the *alanes*, but this mixture was mentioned by all my informants with a necessary comment on the innocence of their relationships with the opposite gender.

Both men and women informants repeatedly downplayed any mention of early sexuality or fancying between boys and girls during their play. Given the strict rules of proven virginity at marriage operating in the area well into the 1960s, perhaps the framing of mixed play groups in an atmosphere of innocence was considered a necessary prologue by my informants. Hopscotch, skipping games, hide and seek, tag, swings, and tossing games such as *mougkos* (similar to the game of duck stone, see Opie and Opie 1997: 84–85) or *abloka* (a game combining throwing and chasing similar to cog stone, see Opie and Opie 1997: 75) and *karydia* (nuts) are cases in point. Some children's games, such as nuts, were part of communal traditions involving more generations. Christmas Day, New Year's Day and the Epiphany were marked by the communal participation of young and old in playing at nuts, a game that retained its festive character of celebration throughout the war.

Some games were gender specific in the sense that they were initiated by boys or girls exclusively, with the occasional participation of a girl marked as a tom-boy or a boy thought to be effeminate or simply easier to manipulate, as in the case of younger boys. Boys' games requiring physical strength were not reported by my informants as being usually attractive to girls, neither were games associated with girlish collections such as scraps or female pretend play scenarios, such as *Koumbares*, mentioned as being attractive to boys. Boys often formed gangs and settled scores by playing at stone wars with sling-shots or bare hands, using the neighborhood creek as a borderland. Stone wars and other forms of ludic warfare, such as exchanging wooden missiles fired from makeshift toy weapons, frequently materialized ethnic antagonisms between adjacent Greek and Armenian settlements, but most Greek male informants downplayed this antagonism in their narratives by adding a hasty remark that 'we continued to hang out together after the battle and were friends as usual.'

Finding the Materials of Play

Cultural historian Bernie Mergen (1992) in his analysis of the changing meaning of toys for US children from 1850 to 1950 mentions that there were three ways of acquiring toys during this period. American children's toys, according to the autobiographical material examined by Mergen, were made, bought, and stolen. This account, while also true of Nikaian children's sources for acquiring playthings during the war period, falls short of mentioning children's work as a resource for finding play materials and children's swopping practices.

Some of the male interviewees mentioned that during the first years of the war, when most materials were scarce (e.g. all clothes were sent to the fighting soldiers), they had to sweep the floors of carpenters and blacksmiths to acquire some coveted material for the construction of complex toys such as scooters. Other makeshift materials were found on the playground of the 'alana' or were sneaked out of kitchens and sewing baskets.

Most of these materials (rags, thread, spools, sticks, broken bicycle wheels, etc.) afforded many play functions, serving as what would currently be termed 'loose parts' (see Almon 2017: 18). Swopping was also a common method of acquiring materials for toy construction. Dimitris A. recalls how he once traded one hundred cigarette box covers for two spare ball bearings he needed for his scooter. In this case, finding the materials of play was part of the play process itself.

Girls' Toys

Most of the makeshift toys that girls played with during the war were used during pretend play. Toys used in girls' make-believe focused on life-cycle rituals and the everyday routines of young married women with children. Marriage marked the transition to the legal status of adulthood for women well before the war (see Papathanasiou 2010 for an account of childhood boundaries in rural areas of the 1930s). During the war, weddings continued despite the famine (Loukos 2007). Girls had a prominent role in wedding preparations and they also enjoyed re-enacting weddings during their street play. One of the girls was dressed as a 'nyfi' (a bride) and a younger boy usually played the bride-groom.

One of the emblematic girls' play themes using a variety of toys and makeshift accessories was *Koumbares*.[4]

Toys Used in Pretend Play: Koumbares

Girls constructed toys for their favorite symbolic play themes, one of which was the game *Koumbares*.[5] Koumbares is a play theme revolving around female friendship cast in a kinship idiom. Female social life in the 1940s in working-class areas was family-oriented and local, and constituted of exchanging home visits. Girls reproduced the rules of hospitality during their Koumbares play sessions by decorating their homes and cooking for the visitors. Identifying with a grown-up female character for some girls meant that they needed a visual cue for their transformation into an adult. Koula remembers how she and her sisters and the girls of their street used to tie a pair of spools (*karoulia*) on their bare feet to act as high heels. The karoulia momentarily transformed the young girls into ladies ready for a Koumbares session.

Other means for transforming girls into ladies—after the famine—were makeshift necklaces and bracelets made of macaroni.

Koumbares sessions also involved playing at cooking, as visitors had to be offered food. Before the war, many girls had a commercial clay pot and a pitcher—possibly sold by peddlers in religious fairs. Koula was one of the ones lucky enough to have this equipment, so she and her friends lit a fire using stones to create friction and prepared meals with grass.

Doll Constructions for Mother-Baby Scenarios

Most female informants played with dolls during their childhood but not all of them during the war actually had a doll of their own, even a rag doll, in which case girls enjoyed dressing and undressing their rag dolls with makeshift dresses during play.

Girls resorted to ad hoc constructions during most of their play-time:

Koula and her sisters borrowed wooden cooking spoons from their mother to make dolls. They first drew the facial features with a pencil, then tied a horizontal stick at the spoon's neck to make hands, and dressed the entire construction with cloth. The spoons were our babies! 'Do you want to eat? Sleep now!' Other girls turned pillows or cushions into temporary dolls.

Girls' Collections: Zografies

Playing at *zografies* (scraps) was another distinctive form of girls' play vividly featuring in the memories of both men and women. Scraps were sold in corner shops in the area and were cheap. Women born in the early- or mid-thirties remember having collections of *zografies* before the war which they had acquired through collecting money from errands to neighbors. Scraps continued to circulate among girls during the war as they were exchanged or gambled during play.

The gambling game played was simple. The girls elected the *mana* (=-mother, the most popular term for power figures and central persons in Greek games) who was responsible for hiding the scraps in the pages of a notebook. Each girl participated in the game by putting one scrap at stake which was hidden by the mana in the pages of the notebook. The mana subsequently allocated one number to each girl and began the game by turning the pages of the notebook counting out loudly 1-2-3-4, etc. The girl whose number coincided with the number of the page where a scrap was found got the scrap.

The game was so absorbing that some female informants recalled delaying to run to air raid shelters because they were immersed in playing at zografies.

Boys also formed collections they integrated in gambling games. Cigarette box covers with logos of Greek—and later English and American—cigarette brands were traded or gambled in a game played like cards.

Boys' Toys and Playthings

Football, marbles, nuts, tops, kites, and hoops featured most commonly as favorite games with things in the elderly men's memories of their boyhoods.

Football has been one the favorite street games for boys for over a century, but rubber or leather balls were not available during the war years and children had to rely on cloth balls made of old socks or stockings. The sock was stuffed with rags and sewn into a round sphere that served the boys as long as it didn't fall into a mud hole. If soaked, the ball was useless, and it took days to dry.

Dynamic Toys[6]

Boys enjoyed toys that could make sounds (such as whistles made of apricot kernels) or move (e.g. yo-yos made of a stuffed rag ball attached to a rubber string) either on water (e.g. paper boats), in the air (kites), or on earth (tops, hoops, spool-shaped tanks, makeshift cars, scooters). Some of these constructions (like the *saita* type of kite and paper boats) were fairly straightforward and simple constructions that even a young boy could make and required minimum material, a piece of newspaper and yarn or string. Hexagon-shaped types of kite and the triangular *tserkeni* type needed more materials and craftsmanship skills, in terms of both finding and assembling the materials.

Paper Boats

One of the main avenues in the area today used to be a creek, before the streets were paved in the 1970s. When heavy rain turned the creek into a torrent, boys would rush out of their homes to play with their paper boats and would spend many hours throwing stones and watching their paper boats travel along the stormy waters. 'There wasn't much to do at the time so we enjoyed simple things like that,' commented Mihalis, who was 6 years old when the war began.

Dimitris made a series of *trata* paper boats, reproducing harbor scenes he had witnessed at the Piraeus port.

Kites

Kites were mentioned as one of the favorite spring pastimes by many informants. The peak of the kite-flying season were the weeks before and after Lent and especially 'Clean Monday,' the first day of Lent, which was a holiday. Three types of kite were most popular in the area: the *saita* (the simplest form of kite without a frame), the *tserkeni* (a triangular kite), and the hexagonal kite. Kites were time-consuming constructions to make, as the boys had to find the reeds for the frame near the creek and then make glue by boiling sugar or flour in water. As sugar and flour were scarce resources during the war, boys were punished for snitching these materials. Not all boys constructed their own kites, however. Those who were more gifted and renowned for their craftsmanship would sell their kites to other boys. Kite stealing, by attaching razors to their own kites which cut the string of a coveted kite in the air, was also widely spread among boys as an act exhibiting their cunning. Kite flying on 'Clean Monday' constituted a form of communal festive play, including many family members and involving more generations (Fig. 11.5).

Fig. 11.5 Making a hexagonal kite at a KAPH centre in Piraeus, for the exhibit "Toys from the Attic", December 2013. Photo by author

Hoops and Scooters

Metal and rubber hoops were played with by both girls and boys, children competing in trundling their hoop as far as possible. Metal hoops required finding the material for the hoop or the trundle guiding the movement, a process involving either running an extra errand for a local ironsmith or, most likely, searching in the *alana* for a spare bicycle rim.

Scooters (*patiniia*) were the most difficult toy to make, as they required a high degree of skill and a lot of scarce materials to construct the frame, the wheels and the handle bar. Two types of scooter were common in the area, the upright *patini* and the airplane-shaped *patini* in which the player lay down. The upright patini served as a means of transport as well as a plaything, while some scooter riders were also reported to help the *saltadoroi* youths at looting German trucks by staging races on the streets frequented by German vehicles. Odysseas, who was one of these kids, explained their method of distracting the enemy. As the truck approached and slowed down in front of a rail crossing, the kids crossed the street on their scooters, causing the German truck to stop. While the German soldiers were occupied in trying to get the scooters out of the way, the s*altadoroi* raided the back of the truck.

Toys for Tossing Games

Marbles and Nuts

Although some women reported that they liked playing at marbles and most of them played at nuts during girlhood years, marbles were considered a boys' game *par excellence.* Boys used either their own makeshift or commercial clay marbles (*voloi*). As self-constructed *voloi* disintegrated more easily, commercial *voloi* had higher value, while the glass marbles (*gyalenia* or *kollenia*) that were scarce in the neighborhood had the highest value of all. *Gyalenia* served as shooters and were acquired through swopping practices, as few children in the neighborhood could afford to buy them. One glass marble was traded for ten clay marbles. Two games with marbles were mentioned as being popular during the war, both highly competitive: the line, and the hole (*lakkaki*), a gambling game similar to odds and evens. Similar games were played with nuts. Shortly after the war, when glassed beverages were available, boys collected and played several tossing games with bottle caps (*kapakia*). Fotis, who was born in 1936, recalls drawing a labyrinth on the ground and competing in reaching the center of the labyrinth with his *kapaki* without touching the contour lines.

Tops

Spinning tops were another competitive form of play among young boys. Some boys like George M. used their fathers' tools to make their own tops, but most boys had commercial tops bought before the famine or won from other players. Tops could either be played with individually for the sake of enjoying spinning, or as a gambling game in which players took turns in trying to knock other tops lying within a circle carved on the ground to beyond the perimeter of the circle. All tops knocked outside the circle were won by the owner of the top which had struck them.

Karaghiozis Shadow Puppets and Performances

Shadow theatre (*karaghiozis*) was a popular spectacle for adults and children for most of the twentieth century, while shadow puppets continue to attract contemporary kids in Greece. Greek children have been reported as having made their own figures and staged makeshift performances in yards during the summer throughout the twentieth century. The most popular themes were those drawing on legendary figures of Ancient Greece, such as Alexander the Great, or on famous warriors of the 1821 Independence War against the Ottoman Turks. *Karaghiozis* performers continued to stage their plays during the German Occupation and their performances were popular among both adults and children; they were also attended by Nazi officers.[7] Boys and girls also continued to play at *Karaghiozis* during the war, charging a small fee to the spectators. Girls collected the money while boys made the figures, scouring the area for materials to make their tools and their figures. White bedsheets served as the screen, while the figures were usually made of cardboard. Children used cutters to carve holes in several points of the figure and covered them with pieces of cloth, to add a touch of color and transparency behind the 'screen.'

Tools for making the shadow puppets had to be improvized. Boys used to place nails on tram rails to flatten the sharp edge and form a *kopidi* (a cutting tool). '*Karaghiozis* the fountain man' and '*Karaghiozis* the baker' were among the popular plays staged by Nikaian children during the war.

How Did the War Influence Play?

As children experienced the sight of starving people and heard stories of dying relatives during the famine, it would be odd if these stories did not appear in their play.

Kaiti made a baby by folding her pillow into a cylinder. Folded in this fashion it reminded her of a swaddled baby, a regular companion for her solitary play sessions. She fed it and put it to sleep but sometimes the baby died and she left it in the yard under a tree. The following day she would collect the baby for new play scenarios, having forgotten that her pillow-baby had died.

The war also inspired scenarios for gun play with self-made wooden guns and bullets for boys and the construction of dynamic toys which moved with the help of makeshift winding mechanisms such as elastic rubber bands. Spools were transformed into tanks by carving notches in the sides of the spool and adding the winding mechanism. Famous battles between the EPON youths and the Nazis in the streets of Nikaia during the last year of the Occupation were an immediate inspiration for young children's war-play scenarios.

Dangerous Play, Play in Times of Danger

Some boys engaged in dangerous play using spent bullets as marbles and bomb shells as targets. If the bullet was unspent, it exploded. Minas A. recalls launching rockets made of cans filled with acetylene, a highly explosive material.

Accidents could happen during play when children carried out their play activities in busy streets. Minas lost his leg when he was run over by a bus while playing with his scooter.

Some published autobiographical material on children's daily life during the war focuses on games in air raid shelters (e.g. Iliou 2012). However, the areas where these recollections originate were situated far from the bombing targets. As the harbor of Piraeus, a regular target of bomb raids by Germans or allies, was close to the neighborhood, my informants recall the horrors of the bombings rather than the relaxed atmosphere of play. The danger was too real for children to relax and play. Yet air raids notwithstanding, most children of both genders learned to control their fear in order to carry on living, playing and helping their family by running errands when they literally crawled their way from one neighborhood to the next to avoid stray bullets. War in this sense was experienced as a dangerous but challenging playground, but none of my informants recalls this period as an era of lost childhood.

Conclusions

One of the problems in dealing with childhood in the past is the fuzziness of boundaries between childhood, youth, and adulthood. Contemporary notions of Western childhood comprise the first 18 years of the life course, a period strongly associated with, among other things, schooling, play, and consumption, rather than with production. Toy play, in particular, marks the boundaries between early and middle childhood, as kids seem to 'grow older at a younger age' with regard to their play preferences (Alfano 2003: 21). Anthropological and historical works on childhood warn us against general or ethno-centric definitions of childhood and claim that the experience of childhood differs according to gender, class, ethnicity, and culture, thus suggesting the existence of multiple childhoods. In this sense, famine and war had different consequences for the children of the poor, as they led less cocooned lives than the children of the middle classes according to historical records on child labor and autobiographical material on childhood memories of the war (see, for instance, Papadopoulos 1995; Romanos 2008).

Social historians dealing with Greek children's experience of WWII have just begun to explore questions regarding the ways childhood was perceived and experienced during this period (e.g. Papathanassiou 2010; Vervenioti 2010). What marked transition from childhood to the category of youth in the Piraeus neighborhoods during the war is still difficult to assess, but play, as the main expression of children's culture, seems most likely to have characterized the primary school years. If a child stopped attending primary school for any reason, she or he would have to contribute to the family budget by seeking regular work. Getting a job then was a step towards gaining the status of youth and later adult, especially in the case of male informants, as adulthood for women was only achieved through marriage.

This is not to say that street play was dropped altogether with the end of childhood, as many games with things were enjoyed by young and old in the days when social life meant sharing leisure time with family and neighbors in the unpaved neighborhood streets and house door steps. George and Mihalis remember playing in the streets on their days off work while Eleni and Vaso, who married young and continued to live with their paternal family for some time after marriage, remember that they still went out to play at scraps and makeshift swings fastened on tree branches or wooden poles.

Toy making formed an integral part of toy play, a time-consuming and innovative bricolage but also a dynamic process uniting work and play, leisure and labor, and promoting children's active participation in the adult world in times of crisis.

Notes

1. See http://www.occupation-memories.org/oral-history/index.htmland https://sites.google.com/site/opiathinas2015/.
2. See Seriff (this volume) for a discussion of the difference between dehumanizing authority figures such as dictators as a strategy of resistance and dehumanizing an entire nation or race as a strategy of planned extinction, as in the case of portrayals of Jews in toys and board games during the Nazi era.
3. It was during the Civil War that teenagers were recruited by both the left wing Democratic Army and the opposed right wing Greek National Army supported by the British and the Americans in 1948–1949. While an expedition to save children younger than 14 was launched by both opponents who evacuated children from warfare sites either to countries of the former Socialist Block or to state operated camps, teenagers over 14 captured as Democratic army fighters were alternatively sent for reform to vocation schools or imprisoned and sent to exile as adults after the defeat of the Democratic Army (Vervenioti 2005, 2010). Definitions of child soldiers vary cross-culturally and historically following variations in perceptions of childhood. Current definitions of child soldiers provided by the UN (see 2002 Optional Protocol to the UN Convention on the Rights of the Child) comprise individuals under 18 years of age serving military purposes in any capacity including that of messenger or look out. The 1949 age limits set by the UN for participation in armed conflict was 15. Here I employ the term 'child soldier' only in the sense of fighter younger than 14 or 15 years of age (see Papathanassiou for 2010 for a discussion of the fuzzy boundaries between childhood and youth during the Interwar period in rural Greece).
4. Koumbares are women related by ***koumbaria*** ties, i.e., ties of spiritual kinship contracted through wedding or baptism sponsorship. Koumbaria ties exist between the wedding sponsor and the married couple and between godparents and the baptized child's natural parents. In the case

of baptism sponsorship, marriage prohibitions are imposed between baptizer and baptized and their linear relatives and between spiritual siblings, i.e., godchildren of the same godparent. While koumbaria ties have traditionally formed the yeast for the development of political patronage in rural Greece (see Campbell 1964) wedding and baptism sponsors in this area were often sought among close relatives, friends and work partners.

5. Koumbares has endured so long as a play theme that it has become synonymous with child's play. Current political discourse often uses the phrase 'Tora tis koumbares the paixoume?' (What is this? Are we playing at Koumbares now?) to declare their serious intentions by juxtaposing play and seriousness. See Gougoulis (2003) for a description of Koumbares sessions in a rural context in the late 1980s and early 1990s.
6. Here I adopt Khannna's classification of Indian folk toys into static and dynamic according to their potential for movement and sound-making. According to Khanna (1983: 3) dynamic folk toys 'provide a sensory experience through their actions: they create movement, change form and make sounds'.
7. Karaghiozis and puppet theatre performers participated in boosting the morale of Anti-Nazi resistance fighters in 'Free Greece' (the mountains of Central Greece controlled by ELAS, (the Greek People's Liberation Army), by staging performances for their children. The themes of these performances drew on the Anti-Fascist struggle (see Velioti-Georgopoulos: 2016). Urban Karaghiozis performers were also inspired by the events of the war and wrote plays such as 'The Gestapo traitor' or 'Karaghiozis the Black Marketeer' (Kiourtsakis 1983: 271–272; Mathiopoulos 2007: 283–284, 286).

References

Abrams, L. (2010). *Oral history theory*. London: Routledge.

Alfano, K. (2003). Making toys for a global market. In L.-E. Berg, A. Nelson & K. Svensson (Eds.), *Toys in educational and socio-cultural contexts*. Stockholm: Sitrek.

Almon, J. (2017). *Playing it up—With loose parts, playpods, and adventure playgrounds*. Annapolis, MD: Alliance for Childhood.

Argyriadi, M. (2008). Τα ελληνικά εμπορικά παιχνίδια κατά τον 20ό αιώνα [Greek commercial toys in the 20th century]. In C. Gougoulis & D. Karakatsani (Eds.), *Το ελληνικό παιχνίδι. Διαδρομές στην ιστορία του. [Greek Toys: Historical trajectories]*. Athens: ELIA and MIET. (In Greek).

Ariès, P. (1973/1960). *Centuries of childhood*. Harmondsworth: Penguin.

Bergen, S., & Niven, B. (Eds.). (2014). *Writing the history of memory*. London: Bloomsbury.

Bournova, E. (2005). Θάνατοι από πείνα. Η Αθήνα το χειμώνα του 1941–1942 [Deaths from hunger. Athens in Winter 1941–1942]. *Arheiotaxio, 7*, 52–73.

Campbell, J. K. (1964). *Honor, family and patronage in rural Greece*. New York and Oxford: Oxford University Press.

Cunningham, H. (1995). *Children and childhood in western society since 1500*. London: Longman.

Desypris, M. (2016). Η κατοχική πείνα στην Νίκαια Αττικής (Κοκκινιά) και η δημογραφική τύχη της πόλης (1941–1944) [Occupation famine in Nikaia, Attica (Kokkinia) and the demographic fate of the city]. In S. Dordanas et al. (Eds.), *Κατοχική Βία 1939–1945. Η ελληνική και ευρωπαϊκή εμπειρία [Occupation Violence 1939–1945. Greek and European experiences]*. Arta and Athens: Municipality of N. Skoufas and Assini Publications.

Eisen, G. (1988). *Children and play in the holocaust. Games among the shadows*. Amherst: University of Massachusetts Press.

Ennew, J. (1986). *The sexual exploitation of children*. London: Polity Press.

Gougoulis, G. C. (2003). *The material culture of children's play: Space, toys and the commoditization of childhood in a Greek community*. Doctoral dissertation, University College London.

Hadjiiossif, C. (2007). Η ελληνική οικονομία, πεδίο μάχης και αντίστασης [Greek Economy, a domain of battle and resistance]. In C. Hadjiiossif & P. Papastratis (Eds.), *Ιστορία της Ελλάδας του 20ού αιώνα [History of 20th Century Greece]* (Vol. 3 [part 2]). Athens: Vivliorama. (In Greek).

Halbwachs, M. (1992). *On collective memory*. Chicago: University of Chicago Press.

Hionidou, V. (2011). *Λιμός και θάνατος στην κατοχική Ελλάδα 1941–1944 [Famine and Death in Occupied Greece 1941–1944]*. Athens: Hestia Publications.

Hirschon, R. (1989). *Heirs of the Greek catastrophe. The social lives of Asia minor refugees in Piraeus*. New York: Berghahn Books.

Honwana, A. (2005). Innocent and guilty. Child soldiers as interstitial & tactical agents. In A. Honwana & F. De Boeck (Eds.), *Makers and breakers: Children & youth in postcolonial Africa*. Oxford: James Currey—Africa World Press—Codestria.

Iliou, M. (2012). *Παιχνίδια πολέμου. Αφηγήσεις*. (*War games. Narratives*). Athens: Hestia. (In Greek).

James, A., & Prout, A. (Eds.). (1990). *Constructing and reconstructing childhood*. London: The Falmer Press.

James, A., Jenks, C., & Prout, A. (Eds.). (1998). *Theorizing childhood*. Oxford: Polity Press.

Khanna, S. (1983). *Dynamic folk toys. Indian play based on the application of simple principles of science and technology.* New Delhi: Office of the Development Commissioner for Handicrafts.

Kiourtsakis, G. (1983). *Προφορική παράδοση και ομαδική δημιουργία. Το παράδειγμα του Καραγκιόζη [Oral Tradition and Collective Creation: the Example of Karaghiozis].* Athens: Kedros publications. (In Greek).

Konstantinou, F. (2007). Γύρω από τη ζωή και το έργο της Βούλας Παπαϊωάννου [On the life and works of Voula Papaioannou]. In F. Konstantinou et al. (Eds.). *Η φωτογράφος Βούλα Παπαϊωάννου [The Photographer Voula Papaioannou].* Athens: Agra Publishers and Benaki Museum. (In Greek).

Kozlovsky, T. (2008). Adventure playgrounds and postwar reconstruction. In M. Gutman & N. de Connink-Smith (Eds.), *Designing modern childhoods: History, space and the material culture of children.* New Brunswick: Rutgers University Press.

Loukos, C. (2007). Η πείνα στην Κατοχή [Famine during the Occupation]. In C. Hadjiiossif & P. Papastratis (Eds.), *Ιστορία της Ελλάδας του 20ού αιώνα [History of 20th Century Greece]* (Vol. 3 [part 2]). Athens: Vivliorama. (In Greek).

Lyberatos, M. (2007). Οι οργανώσεις της Αντίστασης [Resistance Organizations. In C. Hadjiiossif & P. Papastratis (Eds.), *Ιστορία της Ελλάδας του 20ού αιώνα [History of 20th century Greece]* (Vol. 3 [part 2]). Athens: Vivliorama. (In Greek).

Mathiopoulos, E. (2007). Οι εικαστικές τέχνες στην Ελλάδα κατά την περίοδο 1940–1944 [Visual arts in Greece during the period 1940–1944]. In C. Hadjiiossif & P. Papastratis (Eds.), *Ιστορία της Ελλάδας του 20ού αιώνα [History of 20th Century Greece]* (Vol. 3 [part 2]). Athens: Vivliorama. (In Greek).

Mazower, M. (2001). *Inside Hitler's Greece/The experience of occupation 1941–1944.* New Haven: Yale University Press.

Mergen, B. (1992). Made, bought, stolen: Toys and the culture of childhood. In E. West & P. Petrik (Eds.), *Small worlds. Children and adolescents in America. 1850–1950.* Lawrence, Kansas: University of Kansas Press.

Moshenka, G. (2008). A hard rain. Children's Shrapnel collections in the second world war. *Journal of Material Culture, 13,* 107–125.

Opie, P., & Opie, I. (1997). *Children's games with things.* Oxford: Oxford University Press.

Papadopoulos, L. (1995). *Οι παλιοί συμμαθητές (The Old School Mates).* Athens: Kastaniotis. (In Greek).

Papathanassiou, M. (2010). Νεανικές ηλικίες και αγροτική κοινωνία στην Ελλάδα του πρώιμου 20ού αιώνα [Youths in Greek agricultural society in the early 20th century]. In V. Karamanolakis, E. Olympitou & I. Papathanassiou (Eds.), *Η ελληνική νεολαία στον 20ό αιώνα. Πολιτικές διαδρομές, κοινωνικές πρακτικές και πολιτιστικές εκφράσεις [Greek Youth in the 20th Century.*

Political Trajectories, Social Practices and Cultural Expressions]. Athens: Themelio. (In Greek).

Romanos, G. (2008). *Μια αθηναϊκή βεγγέρα (An Athenian Evening Gathering)*. Athens: Potamos. (In Greek).

Seriff, S. (in press). Holocaust war games: Playing with Genocide. In J. Goldstein & L. Magalhaes (Eds.), *Toys and communication*. London: Palgrave.

Sutton-Smith, B. (1986). *Toys as culture*. New York and London: Gardner Press.

Theotokas, G. (n.d.). *Τετράδια ημερολογίου (1939–1944)* (*Diary Notes 1939–1944)*. Athens.

Varon-Vasar, O. (2009). *Η ενηλικίωση μιας γενιάς. Νέοι και νέες στην Κατοχή και την Αντίσταση [A Generation's Coming of Age: Young Men and Women during the Occupation and the Resistance]*. Athens: Estia. (In Greek).

Velioti-Georgopoulos, M. (2016). Playing and learning with puppets. Puppet theatre in 'Free Greece' (1943–1944)'. *Recherches: Culture et histoire dans l' Espace Roman, 16*, 59–65.

Vervenioti, T. (2005). Περί:παιδομαζώματος' και 'παιδοφυλάγματος' ο Λόγος ή τα παιδιά στη δίνη της εμφύλιας διαμάχης. [The discourse on 'child-kidnapping' and 'child-saving' or Children amidst the Greek Civil Conflict]. In E. Voutyra et al.(Eds.), *Το όπλο παρά πόδα. Οι πολιτικοί πρόσφυγες του ελληνικού εμφυλίου πολέμου στην Ανατολική Ευρώπη. [Weapons at the Ready. Greek Civil War Political Refugees in Eastern Europe]*. Thessaloniki: University of Macedonia Press. (In Greek).

Vervenioti, T. (2010). Οι ανήλικοι μαχητές του Δημοκρατικού Στρατού. Από τις φυλακές και τα στρατόπεδα συγκέντρωσης στις Βασιλικές Σχολές Λέρου. [Minor fighters of the Democratic Army. From prisons and concentration camps to the Royal Schools of Leros]. In V. Karamanolakis, E. Olympitou & I. Papathanassiou (Eds.), *Η ελληνική νεολαία στον 20ό αιώνα. Πολιτικές διαδρομές, κοινωνικές πρακτικές και πολιτιστικές εκφράσεις [Greek Youth in the 20th Century. Political Trajectories, Social Practices and Cultural Expressions]*. Athens: Themelio. (In Greek).

West, E., & Petrick, P. (Eds.). (1992). *Small worlds. Children and adolescents in America 1850–1950*. Lawrnce, Kansas: University of Kansas Press.

West, H. (2000). Girls with guns. Narrating the experience of war of Frelimo's 'female detachment'. *Anthropological Quarterly, 74*(4), 180–194.

Author Biography

Cleo Gougoulis is an Assistant Professor of Folk and Popular Culture at the Department of Cultural Heritage Management and New Technologies, University of Patras, Greece, and also Vice President of ITRA.

CHAPTER 12

Can Toy Premiums Induce Healthy Eating?

Carla Ferreira and Luísa Agante

Introduction

Obesity in children is a growing issue in today's society and its rate worldwide has doubled in about 30 years. In 2011, it was estimated that the number of obese children under 5 years old was 40 million worldwide (WHO 2013). Two of the main causes of obesity are the current sedentary lifestyles and unhealthy eating habits.

There are several reasons behind children's unhealthy eating behaviors, marketing activities among them. Among those activities, one that is widely used is toy premiums. According to the FTC (2012), 48 of the biggest companies in the USA spent 393 million dollars on premiums, which ranks 2nd after traditional media.

Despite the extensive research done on some marketing activities Goldberg et al. (1978; Valkenburg and Buijzen 2005; Rexha et al. 2010; Ogba and Johnson 2010), there is little information regarding the offer of toy premiums. Previous studies have examined the effects of toy premiums

C. Ferreira
Nova School of Business and Economics, Lisbon, Portugal
e-mail: carla_ss_fereira@hotmail.com

L. Agante (✉)
School of Economics, University of Porto, Porto, Portugal
e-mail: lagante@gmail.com

L. Magalhães and J. Goldstein (eds.), *Toys and Communication*,
https://doi.org/10.1057/978-1-137-59136-4_12

in children from 4 to 8 years old (Heslop and Ryans 1980) and with children from 6 to 12 years old (Shimp et al. 1976). Both studies found that offering toy premiums paired with breakfast cereals may influence children's preferences. A more recent study (McAlister and Cornwell 2012) with a sample of younger children (3 and 5 years old) investigated the effect of toy premiums on the food choices between unhealthy and healthy meals, and found that toy premiums affect children's attitudes towards healthy and unhealthy food. As far as the authors know, there are no previous studies on the effects of toy premiums on the choice and attitude towards healthy food in older children, and thus there is no evidence of the age at which toy premiums are effective in promoting healthy eating habits. Although it has been shown by Shimp et al. (1976) that toy premiums affect attitudes towards cereals in older children, we wonder if the same effect holds true for healthy food. Hence, among children above 5 years old we expect toy premiums either to be ineffective or to be effective but to a lesser extent than among preschoolers on altering the attitudes and purchase intentions of healthy food.

Literature Review and Hypothesis

Obesity and Children's Food Preferences

In Portugal in 2010, 35.6% of children between 6 and 8 years old were overweight and 14.6% were obese (Rito et al. 2012). Among other causes, children's obesity is caused by increasing unhealthy eating behavior.

There are many reasons that children prefer unhealthy to healthy food. First, unhealthy meals such as fast food are more appealing to children in terms of taste, smell and appearance (Stevenson et al. 2007), and most of the time are easily available when compared with healthy food Shepherd et al. 2006). Furthermore, unhealthy food is associated with friendship and pleasure (Shepherd et al. 2006) and, as children grow, the action of consuming unhealthy meals is seen by children as cool (Schor and Ford 2007). Healthy food is perceived by young consumers as not tasty (McKinley et al. 2005; Stevenson et al. 2007). Recently, many actions have been taken in order to promote healthy eating habits in schools and through other vehicles (Hyland et al. 2006; Ransley et al. 2010), resulting in an increase in the consumption of fruit and vegetables among children from 2005 to 2009 (FTC 2012).

However, young children have fear of tasting new food products—*neophobia*—, which leads them to have less diversified diets (Cooke 2007). Although this has little occurrence among children below 2 years old, it has great influence on children from 2 to 7 years and decreases again from this stage until adulthood (Birch 1999). Hence, it is more difficult for younger children (2–7 years old) to taste new foods and enjoy healthy food. As mentioned before, school-age children are not so influenced by *neophobia* and, therefore, when confronted with healthy options, some children change from unhealthy to more healthy meals (Rexha et al. 2010).

Effects of Marketing on Children's Attitudes and Preferences

Attitude is an opinion about people and things (Solomon et al. 2006). To evaluate a person's attitude towards any object it is necessary to evaluate their feelings, beliefs and intentions towards it (Solomon et al. 2006). Also, consumers' evaluation of an object depends on the beliefs they have about several characteristics of the object. The knowledge about a brand can be inferred from other secondary identities, such as celebrity endorsements and licensing (Keller et al. 2012). Thus, anything paired with a brand, like premiums, is expected to transfer meaning and knowledge about a brand or a product.

Marketing is often claimed to influence the rise of obesity among children and adults since there are many marketing tools that affect children's attitudes and food preferences. Although toy premiums are frequently used as a marketing tool, there are very few studies that have focused on this marketing tool.

Toy Premiums

The use of toy premiums has the objective of attracting children to purchase the company's product by capturing their attention towards the toy being offered. There is a growing trend for offering toy premiums with food products or meals. Many of these toy premiums are collectibles and are widely used to promote unhealthy food. Fast-food advertising (62.5%) contains more collectibles than that of high-sugared breakfast cereals (2.7%) (Page and Brewster 2007). The same trend is followed by branded websites, with 48% of them enclosing collectible products (Henry and Story 2009). In 2009, the biggest quick-service restaurants in the USA such as Burger King and McDonald's spent 341 million dollars on premiums. Breakfast cereals companies accounted for 6.6% of the expenditures on premiums and carbonated beverages with 3%. As expected, none of the 48 companies reported spending money on premiums to promote fruits and vegetables.

In 2009, fast-food restaurants sold around one billion children's meals paired with toy premiums to children under 12 years. In terms of age ranges, fast-food restaurants spent more money on premiums to children between 2 and 11 years old than to children from 12 to 17 years old.

Effects of Toy Premiums on Children

Past research has examined the effect of toy premiums with breakfast cereals in 4- to 8-year-olds (Heslop and Ryans 1980) and in children from 6 to 12 years old (Shimp et al. 1976). Apparently pairing a food product with toy premiums may change children's preferences, although it does not mean a change in children's choices. Toy premiums are seen by parents as a very powerful tool to attract their children (Pettigrew and Roberts 2006).

Concerning the effect of toy premiums on children's brand image, previous literature states that they can increase short-term sales and improve children's brand image (McNeal 1999). However, if the offered premium is unattractive to customers, this may negatively affect the brand image and the attitude toward the brand (Simonson et al. 1994).

McNeal's (1999) conclusions were taken from research with children's food (unhealthy), so we do not know if the same effects can be reached by offering a toy premium with healthy food. In this view, marketing activities could help in improving eating habits of children. In fact, McAlister and Cornwell (2012) explored the reaction of children between 3 and 5 years old to collectible toy premiums with unhealthy and healthy meals. When presented with a healthy food with toy premium and an unhealthy meal without toy premium, preschoolers chose the healthier option. Additionally, it was found that pairing healthy and unhealthy food with toy premiums increased children's attitude towards both types of meal, with the major increase noticed in the healthy food. The same has been done with children from 6 to 12 years old (Hobin et al. 2012) but using well-known toy premiums from McDonald's, which includes the effect of brand familiarity and brand loyalty on the outcomes. In terms of less familiar brands or unknown brands, there is no research stating if this effect also holds true for older children.

We tested the following hypothesis:

> When comparing healthy food with the same food paired with toy premium, children will have a) *a better attitude towards* b) *and a higher purchase intention of* the food with the toy premium.

In younger children, we know that the effect of toy premiums along with food products is different depending on the nature of the toy—namely, we must distinguish collectible toys from non-collectible toys, and collectible toys should be split among superfluous and non-superfluous toys. By superfluous is meant a collectible toy that a child already owns. McAlister and Cornwell (2012) found that preschoolers, when presented with food paired with non-collectible, non-superfluous collectible, and superfluous collectible toy premiums, preferred the food paired with the non-superfluous collectible toy. Surprisingly, for both unhealthy and healthy food, the attitude towards the meals paired with superfluous collectible and paired with non-collectible toy premiums were similar.

Fast-food chains invest large sums of money promoting toy premiums to children from 2 to 11 years old (FTC 2012). However, the type of toy that is paired with the food, which is mainly targeted at children up to 7 years old (Lambert and Mizerski 2011), pertains to characters from movies that target mostly younger children (up to 6 years old). Hence, children in the upper stage of the target audience for these meals may not be attracted by the toys being offered.

Children's Cognitive Development from 7 to 11 Years Old

As we are going to study the effects of toy premiums on children older than 6 years old, we should start by characterizing this age in terms of their cognitive development. According to Piaget, these children are at the concrete operational stage, which contrasts with preschoolers who are considered as pre-operational children. The main difference is that school-age children are able to think logically on the abstract level and analyze simultaneously more than one dimension.

From 6 to 8 years, children become aware that others have different opinions. They are still self-centered, which means that they cannot think from another person's perspective, because this ability is developed only around 8 years of age. From 8 years old, children have the ability to perceive the persuasive intent of advertising, since this requires them to view it from the advertiser's point of view (John 1999). However, they are not able to consider another person's point of view at the same time as their own. Nonetheless, this capacity to think from another person's perspective leads children to recognize the existence of bias in advertising, thus making them skeptical about advertising and less willing to acquire the advertised product (Miller and Busch 1979; John 1999). In addition, older children have more established preferences than younger children, which makes them less receptive to advertising, especially premium-oriented advertising (Heslop and Ryans 1980).

Between preschool and 2nd grade, children begin to make inferences about people based on the products they use. First graders often compare their possessions to those of others in terms of quantity, while older children appreciate objects based on the status that these will give them (John 1999).

Furthermore, previous research discovered that collecting fulfills the need for competition among collectors, who seek to possess more objects than their peers. By the same token, children see collecting as having fun by competing with others (de la Ville et al. 2010). Other reasons behind a child's collection are to escape from boredom and reality, to learn about a certain field, or to satisfy their passion for the objects, and the aspiration to be different from their peers (Baker and Gentry 1996).

As a result of centration—the inaptitude to focus simultaneously on more than one attribute of an object—children until 7 years old are not capable of paying attention to details or comparing objects with precision. Consequently, children before this age accumulate things instead of collecting them (Acuff 1997; John 1999). Accordingly, they value the quantity more than the variety of toys they possess. In opposition, McAlister et al. (2011) found that preschoolers prefer to have one collectible toy (by sharing another toy with another child) rather than two non-collectible toy premiums. By contrast, children in the concrete operational stage already have the capacity to consider several dimensions of a stimulus at a time and are able to analyze objects or brands with more precision (John 1999). Therefore, instead of accumulating, children start to collect. From all of this, we would expect younger children to accumulate toy premiums while older children would be more focused on details and variety.

Our research questions were whether younger and older children evaluate the toy premium differently and therefore will also evaluate differently the healthy food paired with a toy.

And will there be any differences between superfluous collectibles, non-superfluous collectibles, and non-collectible toys alone and paired with healthy food?

Methodology

Pretest

The purpose of our pretest was to select the toys to be used in the main experiment. The toys would have to be appealing to all children aged from 6 to 11 years old. In order to ensure that attitudes were not influenced by brand familiarity, the toys would also have to be new for the children. With the help of a primary school teacher, we started by making a list of possible toys to test and arrived at a short list of possible toys: a bouncing ball, an airplane, three puzzles, two dinosaurs, and three cars.

Afterwards, we tested these toys in order to find the ones that had the same appeal for children of all ages and genders. The pretest was done with seven children (four girls and three boys), from 6 to 10 years old (Mean = 8; Std. Dev. = 1.63).

We gave each child five cards with different smiley faces and the child chose the one that better represented his/her feeling for each of the selected toys. Our goal was to arrive at four toys which would meet the criteria and that would have equal appeal, in order to use them as examples for a non-collectible toy and three toys belonging to the same collection. As a result, a bouncing ball was chosen as the non-collectible toy and three puzzles were chosen as the collectible toys.

Main Study

Participants

The research focused on children from 6 to 11 years old. For the sake of simplicity and to have a cut-off point, we decided to examine only the extremes of the segment but excluded the 1st graders from the analysis since their capacity to read is not fully developed. Additionally, 1st graders are at the same stage of cognitive development as preschoolers, the pre-operational stage, which had already been studied in previous literature. Hence, the study focused only on 2nd and 4th graders, which corresponds to children in the concrete operational stage of cognitive development.

The sample was composed of 106 children (44.8% boys), from 6 to 11 years old (Mean = 8.08 and Std. Dev. = 1.182), of whom 56 children were from the 2nd grade and 50 were 4th graders. Participants were recruited from schools in the metropolitan area of Lisbon.

Procedure

The objective of the study was to evaluate the effects of having a toy premium paired with healthy food on the attitude towards healthy food. This attitude would be dependent on the attitude towards the toy, which we assumed would decrease with age and would depend on the type of premium (collectible vs. non-collectible).

Therefore, the sample was divided into three groups, two control groups and one experimental group. Both control groups served the purpose of evaluating separately either the healthy food (group F) or the toy (group T) without pairing both, while the experimental group (group E) evaluated the pair food/toy. Table 12.1 summarizes the division per sub-group.

Each group was presented with a picture of the respective item(s) to be evaluated (food, toy, food + toy) and was asked to evaluate their attitudes

Table 12.1 Research groups

Groups		*Experimental conditions*		*Number of children*	*Percentage (%)*
		Healthy food	*Toy premium*		
Group F		Yes	No	20	18.9
Group T	T1	No	Non-collectible	9	8.5
	T2	No	Collectible 1	7	6.6
	T3	No	Collectible 2	10	9.4
	T4	No	Collectible 3	10	9.4
Group E	E1	Yes	Non-collectible	19	17.9
	E2	Yes	Collectible non-superfluous	16	15.1
	E3	Yes	Collectible superfluous	15	14.2
Total				106	100.0

Fig. 12.1 Image questionnaire F, image compilation from three different websites: http://www.baressp.com.br/bares/agua-doce-cachacaria-moema, http://recipetov.net/vegetable-soup-recipe-food-network/, https://marquinhosribeiro.wordpress.com/2012/09/20/um-copo-de-leite/

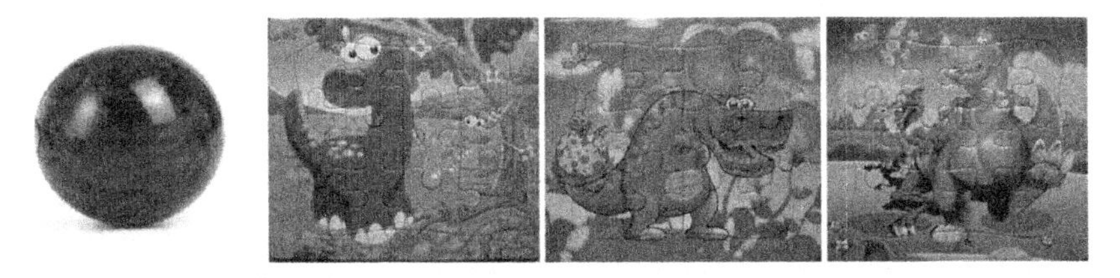

Fig. 12.2 Images questionnaires T1, T2, T3 and T4 http://www.ooops.es/?product=bola-para-rolla-bolla

toward the items in the picture. Finally, in order to assess children's and educators' eating habits a questionnaire was given to educators. (Figs. 12.1 and 12.2)

Measures

Children's attitude towards the healthy food was measured by asking them how much they liked the food and how good it seemed to taste

(McAlister and Cornwell 2012). Additionally, purchase intention was measured by asking children if they would like to buy or ask their parents to buy the food (Phelps and Hoy 1996). Both attitudes and purchase intentions were measured using a five-point smiley faces scale. Attitude towards the toy premium was measured with a five-item scale adapted from several authors (Shimp et al. 1976; Pecheux and Derbaix 1999; Osgood et al. 1957). Children were asked how much they liked the toy and how much fun, cool, and pretty it was. They were also asked how much quality the toy had. All scales were reviewed by a child psychologist in order to evaluate their suitability for children of this age.

Finally, to assess educators' and children's eating habits (Dixon et al. 2007), educators were asked to state their and their child's weekly consumption of vegetables, fruits, French fries, sweets, and soft drinks (Elfhag et al. 2008). They were also asked to rate their and their child's eating habits in terms of healthiness.

Results

We started by analyzing our main hypothesis that, when *comparing healthy food alone with the same food paired with a toy premium*, children would have a better attitude towards and would have a greater purchase intention of the food with the toy premium.

Concerning the effects of pairing healthy food with toy premiums, it was found that pairing the food with the non-collectible toy, a bouncing ball, did not lead to significant changes in the attitude towards healthy food nor in the children's purchase intention ($p > 0.05$). Pairing the healthy food with the three collectible toys did not lead to significant changes in the attitude towards healthy food ($p > 0.050$) but the changes in purchase intention were ambiguous (p(t-test) = 0.422; P(LR) = 0.054 < 0.100). [LR stands for Likelihood Ratio.] Further, pairing healthy food with superfluous collectibles lowered children's attitude towards the food but not in a significant way ($p > 0.05$).

Additionally, comparing the results from healthy food paired with non-superfluous collectibles and paired with superfluous collectibles, the purchase intention ($p = 0.147$) and likability for healthy food did not suffer significant changes ($p = 0.418$). As exception, the changes in anticipated taste were significant according to the non-parametric tests (p(LR) = 0.048) but non-significant on the parametric tests (p(t-test) = 0.884). Thus, we reject the hypothesis.

Since our hypothesis was not confirmed, we proceeded to our research question with low expectations. There was no significant relationship between likability ($p = 0.313$), anticipated taste ($p = 0.183$), and purchase intention ($p = 0.564$) and age. Thus, it was not proven that younger children have greater attitudes and purchase intentions for healthy food with toy premiums than older children.

Regarding children's attitude towards toys alone, it was found that there is a significant difference between likability among 2nd and 4th graders with the likability for the toys being negatively related with age. The same results are valid for opinions about the toys' fun aspect, its beauty, coolness, and quality. With respect to non-collectibles paired with healthy food, no relationship between attitude towards the toy and age was found, and neither were significant differences on attitude between 2nd and 4th graders, with the exception of likability for the toy ($p < 0.100$), which was greater for older children. When pairing collectible toys with healthy food no relationship was found between children's attitude toward the toys and age, with the exception of quality, which was negatively related to age ($p = 0.041$). No significant differences on attitude towards collectibles were found between younger and older children, with the exception of beauty and quality ($p < 0.100$), which were higher for 2nd graders. Once again, when pairing superfluous collectibles with healthy food no relationship between attitude towards the toys and age emerged. No relationship was found between the attitude towards and purchase intention for healthy food with age when the food was paired with the non-collectible. Further, no significant differences occurred on attitude towards healthy food between 2nd and 4th graders. The same results were obtained when healthy food was paired with collectible toys. On the contrary, pairing healthy food with superfluous collectibles led to lower attitude towards healthy food by 2nd graders than by 4th graders ($p = 0.054 < 0.1$). Nonetheless, purchase intention among younger and older children did not register significant differences. In addition, a positive relationship was found between likability for healthy food and age when the food was offered along with superfluous collectibles ($p = 0.030$), but no relationship between anticipated taste and purchase intention with age emerged.

It was also found that there are no differences between attitude towards superfluous collectibles and attitude towards non-collectibles among 2nd graders, with the exception of likability of the toy, which registered only a slight increase ($p = 0.052 < 0.100$) from non-collectibles to superfluous collectibles. In the same direction, no significant differences between superfluous collectibles and non-collectibles were found among 4th graders on all items except fun, which revealed ambiguous changes in terms of significance. Furthermore, the pairing of healthy food with the non-collectible toy did not lead to significant changes on attitude towards the food among 2nd and 4th graders separately. The same happened with the introduction of the three collectible toys, with the exception of purchase intention by 2nd graders, which was ambiguous in significance (P(t-test) = 0.270; p (LR) = 0.076 < 0.100). With the introduction of the superfluous collectibles, no significant changes arose on attitude towards healthy food, with the exception of the likability of the food by 2nd graders, which was lower in the group with the toy (p(t-test) = 0.743; p(LR) = 0.047). Comparing the results from healthy food paired with non-superfluous and superfluous collectibles, there was a higher anticipated taste when the food was paired with superfluous collectibles among 4th graders, but this difference is inconclusive in terms of significance. Nevertheless, among the 2nd graders, it is clear that no significant changes occurred.

Comparing the attitudes towards healthy food paired with non-collectibles and paired with superfluous collectibles, no significant differences were found among 2nd graders. Non-collectibles were associated with higher likability but lower anticipated taste and purchase intention, though these differences were not significant ($p > 0.050$). Moreover, among older children, no significant differences arose. Non-collectibles were associated with lower likability and anticipated taste and greater purchase intention, though these differences were not significant ($p > 0.050$).

Discussion

One of the main conclusions of the present research is that, in general, pairing healthy food with toy premiums does not affect 6- to 11-year-old children's attitude towards and purchase intention for healthy food. In fact, and contrary to McAllister and Cornwell's (2012) discovery, the introduction of any type of toy (non-collectible, collectible, and superfluous collectible) was not effective in increasing children's attitude towards and purchase intention for healthy food. One reason for these results may be the already positive attitude towards healthy food of children when food is presented alone. Thus, although the introduction of toy premiums alters children's attitude towards the food, this difference is not significant. Comparing with McAlister and Cornwell's (2012) study, the contradictory results may be due to the difference in eating habits between the USA and

Portugal. In fact, the children that participated in our study and their parents reported high levels of consumption of fruits and vegetables and low levels of consumption of French fries, sweets, and soda. Although the study evaluated the reactions of children in the short run, a second reason may be linked to the decrease of motivation in the long run when a likable food is paired with a reward (Birch et al. 1982, 1984). Another reason for the ineffectiveness of toy premiums to increase children's choice and attitude towards healthy food may be the comprehension of the persuasive intent of the offer.

A deeper investigation enabled us to conclude that the attitude towards and purchase intention for healthy food were not related to age. Further, as children grow older, they have more positive attitudes toward toys alone, which was reflected in the differences between younger and older children. Also, younger and older children had similar attitudes towards non-collectibles, collectibles, and superfluous collectibles. Because of this, no relevant differences on attitude towards and purchase intention for healthy food paired with the toys were found. The exception happened when younger children had lower attitude towards healthy food than older children but similar purchase intention if the food was paired with superfluous collectibles. One reason might be that pairing healthy food with superfluous collectibles had negative effects on the 2nd graders' opinions about coolness and quality of toys and on the 4th graders' opinion about quality. Additionally, younger as well as older children have similar attitudes toward non-collectibles and superfluous collectibles, as shown by McAllister and Cornwell (2012). The divergence occurred on the likability for the toys among 2nd graders, who liked non-collectibles less than superfluous collectibles.

Limitations and Further Research

One of the main limitations of the study is the sample size, which did not allow the stipulated margin of error in some of the groups. The reduced size of the sample did not allow the effects of each collectible toy to be analyzed separately. Hence, future research should incorporate a larger sample in order to draw more certain conclusions. Secondly, children's stated enjoyment of healthy food may not match their real opinions when faced with the actual food. However, children's consumption of healthy food reported by parents matches with children's liking of healthy food. Besides, during the individual interviews it became clear that children enjoy healthy food in general. Nonetheless, it is possible that those reports of both children's and parents' eating habits are biased by social desirability. Moreover, it is possible that, when presenting healthy and unhealthy food to the children, the choice and attitude towards healthy food present different results. Thirdly, the anticipated taste stated by children may be different from the actual taste when trying the food. Fourth, a different assortment of toys, perhaps more appealing to this age group, might reveal a different pattern of results.

Moreover, the participants in the study were aged between 6 and 11. In countries with high levels of healthy food consumption, the results for children under 6 years old may be different. Furthermore, this research did not evaluate differences between genders. It is expected that girls have a more positive attitude towards healthy food when presented alone (Levin and Levin 2010; Hobin et al. 2012). Finally, further research should seek an explanation for the similar attitudes of children towards non-collectibles and superfluous collectibles.

References

Acuff, D. (1997). *What Kids Buy and Why.* The Free Press.

Baker, S. M., & Gentry, J. W. (1996). Kids as collectors: A phenomenological study of first and fifth graders. *Advances in Consumer Research, 23*(1), 132–137.

Birch, L. L. (1999). Development of food preferences. *Annual Review of Nutrition, 19*(1), 41–62.

Birch, L. L., Birch, D. Marlin, & Kramer, L. (1982). Effects of instrumental eating on children's food preferences. *Appetite, 3*(2), 125–134.

Birch, L. L., Marlin, D. W., & Rotter, J. (1984). Eating as the 'Means' activity in a contingency: Effects on young children's food preference. *Child Development, 55*(43), 1–439.

Cooke, L. (2007). The importance of exposure for healthy eating in childhood: A review. *Journal of Human Nutrition & Dietetics, 20*(4), 294–301.

Dixon, H. G., Scully, M. L., Wakefield, M. A., White, V. M., & Crawford, D. A. (2007). The effects of television advertisements for junk food versus nutritious food on children's food attitudes and preferences. *Social Science and Medicine, 65*(7), 1311–1323.

Elfhag, K., Tholin, S., & Rasmussen, F. (2008). Consumption of fruit, vegetables, sweets and soft drinks are associated with psychological dimensions of eating behaviour in parents and their 12-year-old children. *Public Health Nutrition, 11* (9), 914–923.

Federal Trade Commission FTC. (2012). https://www.ftc.gov/sites/default/files/documents/reports/review-food-marketing-children-and-adolescents-follow-report/121221foodmarketingreport.pdf. Accessed 7 Nov 2013.

Goldberg, M. E., Gorn, G. J., & Gibson, W. (1978). TV messages for snack and breakfast foods: Do they influence children's preferences? *Journal of Consumer Research, 5*(2), 73–81.

Henry, A. E., & Story, M. (2009). Food and beverage brands that market to children and adolescents on the internet: A content analysis of branded web sites. *Journal of Nutrition Education and Behavior, 41*(5), 353–359.

Heslop, L. A., & Ryans, A. B. (1980). A second look at children and the advertising of premiums. *Journal of Consumer Research, 6*(4), 414–420.

Hobin, E. P., Hammond, D. G., Daniel, S., Hanning, R. M., & Manske, S. (2012). The happy meal® effect: The impact of toy premiums on healthy eating among children in Ontario, Canada. *Canadian Journal of Public Health, 103* (4), 244–248.

Hyland, R., Stacy, R., Adamson, A., & Moynihan, P. (2006). Nutrition-related health promotion through an after-school project: The responses of children and their families. *Social Science and Medicine, 62*(3), 758–768.

http://www.baressp.com.br/bares/agua-doce-cachacaria-moema. Accessed 7 Nov 2013.

http://recipetov.net/vegetable-soup-recipe-food-network/. Accessed 7 Nov 2013.

https://marquinhosribeiro.wordpress.com/2012/09/20/um-copo-de-leite/. Accessed 7 Nov 2013.

http://www.ooops.es/?product=bola-para-rolla-bolla. Accessed 19 Oct 2013.

John, D. R. (1999). Consumer socialization of children: A retrospective look at twenty-five years of research. *Journal of Consumer Research, 26*(3), 183–213.

Keller, K. L., Apéria, T., & Georgson, M. (2012). *Strategic brand management: A European perspective* (2nd ed.). England: Pearson Education Limited.

Lambert, C., & Mizerski, R. (2011). *Kids, toys and fast food: An unhealthy mix?* ECU Publications.

Levin, A. M., & Levin, I. P. (2010). Packaging of healthy and unhealthy food products for children and parents: The relative influence of licensed characters and brand names. *Journal of Consumer Behaviour, 9*(5), 393–402.

McAlister, A. R., & Cornwell, T. B. (2012). Collectible toys as marketing tools: Understanding preschool children's responses to foods paired with premiums. *Journal of Public Policy & Marketing, 31*(2), 195–205.

McAlister, A. R., Cornwell, T. B., & Cornain, E. K. (2011). Collectible toys and decisions to share: i will gift you one to expand my set. *British Journal of Developmental Psychology, 29*(1), 1–17.

McKinley, M. C., Lowis, C., Robson, P. J., Wallace, J. M., Morrissey, M., Moran, A., et al. (2005). It's good to talk: Children's views on food and nutrition. *European Journal of Clinical Nutrition, 59*(4), 542–551.

McNeal, J. U. (1999). *The kids market: Myths and realities.* Ithaca, NY: Paramount Market Publishing.

Miller, J. H., Jr., & Busch, P. (1979). Host selling vs. premium TV commercials: An experimental evaluation of their influence on children. *Journal of Marketing Research, 16*(3), 323–332.

Ogba, I.-E., & Johnson, R. (2010). How packaging affects the product preferences of children and the buyer behaviour of their parents in the food industry. *Young Consumers: Insight and Ideas for Responsible Marketers, 11*(1), 77–89.

Osgood, C. E., Suci, G., & Tannenbaum, P. (1957). *The measurement of meaning.* University of Illinois Press. In Bruner, G. C., Hensel, P. J., & James, K. E. (1992). *Marketing scales handbook: A compilation of multi-item measures.* Chicago, IL: American Marketing Association.

Page, R. M., & Brewster, A. (2007). Frequency of promotional strategies and attention elements in children's food commercials during children's programming blocks on US broadcast networks. *Young Consumers: Insight and Ideas for Responsible Marketers, 8*(3), 184–196.

Pecheux, C., & Derbaix, C. (1999). Children and attitude toward the brand: A new measurement scale. *Journal of Advertising Research, 39*(4), 19–27.

Pettigrew, S., & Roberts, M. (2006). Mothers' attitudes towards toys as fast food premiums. *Young Consumers: Insight and Ideas for Responsible Marketers, 7*(4), 60–67.

Phelps, J., & Hoy, M. (1996). The Aad-Ab-PI relationship in children: The impact of brand familiarity and measurement timing. *Psychology & Marketing, 13*(1), 77–105.

Ransley, J. K., Taylor, E. F., Radwan, Y., Kitchen, M. S., Greenwood, D. C., & Cade, J. E. (2010). Does nutrition education in primary schools make a difference to children's fruit and vegetable consumption? *Public Health Nutrition, 13*(11), 1898–1904.

Rexha, D., Mizerski, K., & Mizerski, D. (2010). The effect of availability, point of purchase advertising, and sampling on children's first independent food purchases. *Journal of Promotion Management, 16*, 148–166.

Rito, A. I., Paixão, E., Carvalho, M. A., & Ramos, C. (2012). *Childhood obesity surveillance initiative: COSI Portugal 2010.* Lisbon: INSA, IP.

Schor, J. B., & Ford, M. (2007). From tastes great to cool: Children's food marketing and the rise of the symbolic. *The Journal of Law, Medicine & Ethics, 35*(1), 10–21.

Shepherd, J., Harden, A., Rees, R., Brunton, G., Garcia, J., Oliver, S., et al. (2006). Young people and healthy eating: A systematic review of research on barriers and facilitators. *Health Education Research, 21*(2), 239–257.

Shimp, T. A., Dyer, R. F., & Divita, V. (1976). An experimental test of the harmful effects of premium-oriented commercials on children. *Journal of Consumer Research, 3*(1), 1–11.

Simonson, I., Carmon, Z., & O'Curry, S. (1994). Experimental evidence on the negative effect of product features and sales promotions on brand choice. *Marketing Science, 13*(1), 23.

Solomon, M., Bamossy, G., Askegaard, S., & Hogg, M. K. (2006). *Consumer behaviour: A European perspective.* 3rd ed. Prentice Hall.

Stevenson, C., Doherty, G., & Barnett, J. (2007). Adolescents' views of food and eating: Identifying barriers to healthy eating. *Journal of Adolescence, 30*(3), 417–434.

Valkenburg, P. M., & Buijzen, M. (2005). Identifying determinants of young children's brand awareness: Television, parents and peers. *Journal of Applied Developmental Psychology, 26*(4), 456–468.

de la Ville, V.-I., Brougère, G., & Boireau, N. (2010). How can food become fun? Exploring and testing possibilities. *Young Consumers: Insight and Ideas for Responsible Marketers, 11*(2), 117–130.

WHO. (2013). World Health Organization. http://www.who.int/mediacentre/factsheets/fs311/en/. Accessed 26 Sep 2013.

Author Biographies

Carla Ferreira has a Master's in Management from the Nova School of Business and Economics, Portugal.

Luísa Agante (Marketing Ph.D., ISCTE-IUL) is Professor of Marketing at University of Porto, School of Economics, Portugal.

CHAPTER 13

You Are What You Eat: Toying with the Process of Becoming

Mariah Wade

Introduction

Toys are longitudinal experiential artefacts. For children and youngsters, toys are multi-faceted instruments of cognition and action, knowledge transmission and acquisition: they carry messages and mediate experiences. Toys also communicate the past to the present as memory indices and they work as affect conduits, replete with snapshots, gestures, odors, and tastes, material objects that act as sleights-of-hand of the sensory. Toys and food are intimately connected with selfhood and self-identification as children's toys mediate the practices of food consumption, preparation, and commensality, following local cultural patterns and traditions. These early childhood experiences and the habitual reinforcement through play-acting prime the senses to identify with foods, tastes, and smells that feel and signify 'home' and comfort. Displaced peoples, particularly adults, carry this backpack of yearnings: a craving for foods that is an essential longing for the shelter of home place.

If toys in the plural are all that is said above, a toy is any object that engages and is engaged by a child or an adult and morphs into something else in the process of playing. A toy is a stick, a pebble, the cap of a bottle,

M. Wade (✉)
Department of Anthropology, University of Texas, Austin, TX, USA
e-mail: m.wade@austin.utexas.edu

L. Magalhães and J. Goldstein (eds.), *Toys and Communication*,
https://doi.org/10.1057/978-1-137-59136-4_13

the lid of a pot, a drawing in a book, an embroidered doily, and pasta letters in a bowl of soup. A toy is anything material that can be toyed with, or immaterial as in toying with an idea or a memory. Objects and ideas, or objects/ideas leap out of their material and immaterial boundaries in the act of playing, and they emerge into the realm of the imaginary only to revert to their inert condition as playing ceases. Just like a puppet, there are strings attached to the 'making' of a toy: the actor's play and make-believe must breathe magic into the prosaic.

This chapter explores notions of being and becoming through the materiality of toys and food and the immateriality of the states of being, feelings, and affects engendered by both toys and food in the experiential and habitual practices of users in the past and in the present. This research uses autobiographical moments, conversations, observations, and comparative materials to discuss how play and food practices inculcate habits and substantiate traditions, while at the same time trapping subjects in those same habits and traditions. The entanglement that results from the connection between toy and food make-believe practices is particularly relevant for immigrants as the first generation will emphasize those traditional practices to maintain homeland values and traditions, while posterior generations seek to play out those traditions to *support* hybridity as a cosmopolitan value to be called on in the country of their birth, but principally in the birth country of their parents. Ultimately, I aim 'to show how the things that people make, make people' (ed. Miller 2005: 38), even when we dismiss or forget the object behind the subject.

Reflexive

1950s

Everything mother did for the child was marked by stories: penguins jumping rope on a sunny day spread over a skirt border, bees hopscotching from flower to flower on a cardigan, the indefatigable ant that worked all summer to get food for the winter on a jumper, an embodied object lesson, or folk dancers who became alive on the pockets of a summer dress as the girl twirled. Sometimes, mother embroidered doilies for the child that told stories about food. Embroidered clothing and doilies are toys, as seen in Fig. 13.1.

There is a routine. The mother lays out the rag rug on the wooden floor of the living room. From the kitchen, mother will keep an eye on the child.

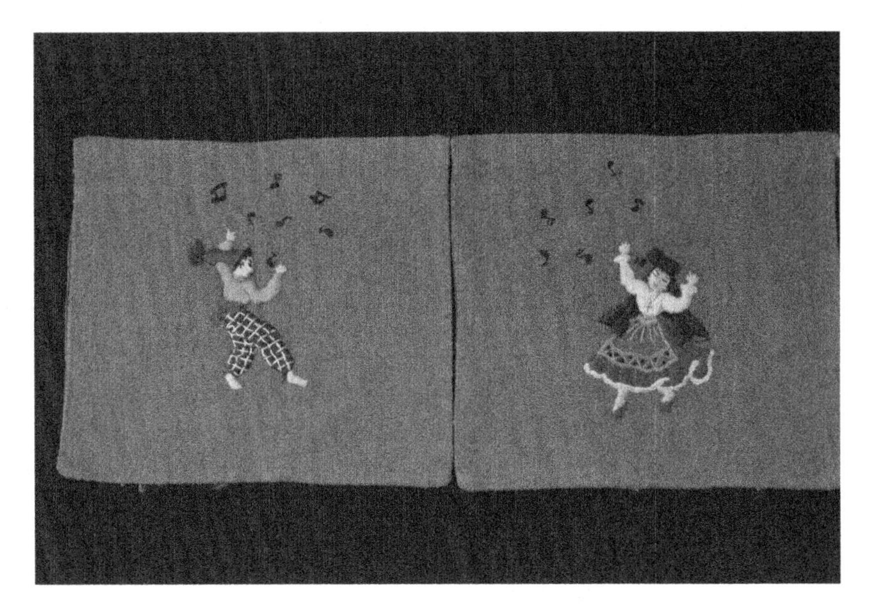

Fig. 13.1 Embroidered *dancers,* owned by author

She tells the child she has to place all her toys and play within the area of the rug—the rug is the child's house and playground. The child will obey the rules strictly. There is a relationship between the objects and the subject: little, clumsy fingers wrapped around small delicate toys, modeling habits and socializing the child. 'We were brought up with the expectations characteristic of our particular social group largely through what we learn in our engagement with the relationships found between everyday things' (ed. Miller 2005: 6) and later, the immateriality of the sensory is fingered through the remembered thing. 'We cannot know who we are, or become what we are, except by looking in a material mirror' (ed. Miller 2005: 8). There is a routine.

From the rug, hunched over, cross-legged, the child eyes the mother and her movements. Slightly bended to the right, hand on the hip, wooden spoon in hand, mother stirs the soup. Mother cooks well but does not like cooking; she would rather be reading. But the child is a picky eater and the chicken soup is a ritual. Pasta letters float in a translucent broth: one bright yellow chicken egg yolk to start, one to get the child to the middle of the soup, and another to finish. That is the bribe. The pasta alphabet will provide the words for the story the mother makes up as she goes along. Selecting the letters is part of the ritual and guides the story. A 'd' for the dog they saw on the street, a 'w' for the color of the dog chasing the birds on the square near their apartment, a 'b' for the birds ... and so the story goes, stimulated by the promise of the egg yolks the child adores. The eye catches the pasta letters as they float and offer endless possibilities before being consumed. Even catching them is a game that nurtures skills. Mother is a storyteller and a domestic toy inventor. Soup is a toy. 'It is good to *think* in stories' (Block 2006: 6), and the pasta letters provide food for thought and are open-ended to story-telling. Mother embroiders life experiences into place through verbal story-telling, such as she does with needle and thread.

Within the marked boundaries of her rug-house, the child will prepare meals in her kitchen toys. A parsley leaf, a potato peeling, a shred of apple, a drop of water all mashed down and placed on the colorful tin-sheet stove, as in Fig. 13.2. She will carefully watch and mimic mother's moves,

Fig. 13.2 Tin-sheet stove, owned by author

attentively performing and learning. She will 'feed' her dolls, each rhythmic miniature spoonful followed by muttered bits of a story, and she will punctuate her role-play, repeating in a stern voice mother's mantra: 'Don't drop crumbs on the floor.'

She will keep her best china tea set for special occasions, the birthday party of her favorite doll. At holidays and birthday parties she has learned that displays matter. She has learned to select the appropriate materials for different occasions: aluminum pots are appropriate for cooking, but the fine porcelain sets must be used for guests and unique occasions. This choice marks the distinctiveness of the event. She has also learned the care she must take with her ceramic sets: she must wash and handle them carefully to avoid breakage. They must last as do those her mother uses. She is learning what her culture wants her to learn, through toys and food. The child has no television or any other media to serve as models, and seldom does she observe other children of her age playing. She plays alone and her learning model is her mother and sometimes her aunts. She learns by observation and mimicry. She is a child of another century when toys were the media for performance within the world of the family (Sutton-Smith 1986: 21–24) and most children did not hold parties of their own. Toys were active objects in the formation of the child-subject whose small body and absorbing mind actively strove to overcome the awkwardness of physical movements while internalizing cultural norms (ed. Miller 2005: 18). Play time and play space constituted special occasions (coordinates), during which the child asked her mother endless questions and quite often sought her mother's opinion as to what dress the doll should wear or what dishes to use. Sometimes mother got tired of the child's follow-up 'whys?' but that was the process of acquiring and archiving information, minutiae that will characterize the adult's capacity for detail and memorization (Sutton-Smith 1986: 25).

The child will travel and live in many countries. She will become a person of many cultures, a hydra with many heads where languages travel through different cultural tracks as in a railway hub: an alphabet soup. These 'other' mimetic learnings will layer as depositional cultural strata and she will call on them to blend in as she sheds cocoons only to be enveloped by others. Her whole will be made of patched-up fragments, time and space specific. But there is slippage in the self, though she will learn to use these habitus fractions to her own advantage: blame it on being 'foreigner' (Adams 2007: 46–47). She has become her own 'self-reflexive project' (Giddens 1991: 32).

She was never an immigrant. As she moved through countries and continents, she lost her 'place' in the game board of countries; that was just it—in the process of becoming, she feels she belongs nowhere. To put it another way, she is, and always will be, 'other' to all 'others' and that seems to vacate 'otherness.' Nevertheless, to create a space of belonging she will mimic her own country foods, but the composite she prepares falls short in smells and tastes, if not ingredients. She will dream of her foods instead. It is the immateriality of her wants that confers expression on her belonging. She can always claim nothing is like what she ate when she was growing up or what one can eat 'back home.' The sensory discord between any version of the foods in the present and those foods her memory recalls sustain the dialogic relationship between the materiality of past foods and the immateriality of her desire for them and the secret wish they remain unequaled (Buchli 2016: 5; Miller 2005: 28).

All the expatriates she knows experience intense food cravings and dream about homeland foods. Invariably, when they meet, the conversation turns to the traditional dishes they miss; an alphabet of cravings. Belonging is translated in tastes and smells unique to growing up, the sensory grip of shared places and dishes. Working late, as she is doing now, she will wish for traditional snacks that can mitigate the silence and the night. As her mind moves from room to room in the house she grew up and she works through her imaginary, she wonders if she sees the silhouette of her mother standing against the kitchen stove or if she imagines it, a memory sieved through her desire to see and to merge disparate worlds (Trend 2013: 4–5).

But, 'playing house is not meant to be real house playing' (Sutton-Smith 1986: 139). Her memory is channeled through the objects —the china sets, the doilies and the 'missing toys' like the red telephone that displayed the Red Riding Hood story as you dialed. She is constituted by hits and misses. She will keep some toys and, toying with them, she will remember her rug-house as she dwells in immigrant houses; affects will be effects, or immateriality cum materiality.

Girl at Play

2010

As seen in Fig. 13.3, the girl stands in front of her kitchen center, an apron to protect her clothing, and prepares tea and sweets for her friends.

Fig. 13.3 Girl playing at cooking meals in her mother's role, author's own photograph

Though the kitchen center is modern, the miniature china is antique. The girl requested the kitchen center as a Christmas present and mother added the china. As Pablo Neruda stated, 'Everything is ceremony in the wild garden of childhood' (1986: 45). For the girl, this is a ceremony of self-hood, performance and commensality. Intrinsically, she knows the meaning of all those concepts and their translation in materiality. She knows that the performance of commensality and display of her skills as a cook and hostess are essential to her status among her friends and schoolmates. The toys are the means to define and legitimize her situated personhood and proclaim her gendered aptitudes. As Brown suggested, 'miniaturization allows [the girl] to grasp totality' (1998: 949) and ground herself in the social milieu, as she stumbles through the process of becoming.

She is a girl from the village of Caxinas located between the towns of Vila do Conde and Póvoa de Varzim in the north of Portugal. Caxinas is a tight community of fishermen families and they have specific traditions to uphold. As a Caxineira, she is already Portuguese, not the other way around—that is her identity. At 6 years old, the girl does not much care for dolls, unless they are virtual babies, soft, passive, round-faced, big-eyed and tiny-mouthed—just big enough that she can bottle-feed the 'baby,' bathe him, and change his diapers. The girl seeks the maternal skills-set; play-practice. She wants to imitate what she thinks her mother would have done with her, and with her brother, in these solitary baby doll interactions. These are 'mother-doll' private attachment moments during which toy/play and food/nurturing are bond (Sutton-Smith 1986: 45). A recent advertisement from Corolle shows a Caucasian, an Asian, and a Black doll, and states: 'Just Like Mommy: Dolls Feat. Make-believe fun gets real' (Corolle 2015). The girl's baby doll is cute, the imagined generic, perfect baby marked as vulnerable, and more important, silent (Ngai 2005). The passive doll will serve as proxy for a baby's bodily and nurturing needs and the girl will mouth noises of contentment or displeasure as she rocks the doll in her arms.

The girl's baby doll is genderless, though the girl will make it male and baptize it accordingly. The naming ceremony signals possession, domestication, and control, and the girl invites her friends to the feast. This too is a rehearsal that asserts capabilities and social status in which food will have a prominent role. Mother will prepare the *real* food but the girl will also serve the tea and some mini-sweets in her toy sets (soda will replace most of the tea drinking). The party is an opportunity to display the toys and gauge how they measure against those of her neighbors and schoolmates. The doll's capabilities are assessed carefully, but it is the girl who is being evaluated for her success in convincing her mother to buy the baby doll, as well as the girl's mastery at managing the show-and-tell of what the doll can do and what the girl can do with the doll (Russ 1993: 33–34). It is a test. She has to prove her skills to her schoolmates but also to her mother. She has rehearsed. Object and subject, as interactive agents, craft the girl's self as she maneuvers to perform as she conceives her mother would.

Recall the Corolle advertisement mentioned above: make-believe fun gets real. Realization of the self *is* a product of practical activity, in as much as the practices in making-believe have the potential to be real because neither has a separate existence for the subject (Rowlands 2005: 80–81). But 'the nurturance of agency is a delicate balancing act' in this make-believe-real interaction (Trend 2013: 28). As the child juggles the different audiences and practices self-monitoring, too much structure might stifle the child's capacity to err and understand the need for self-control, while too much freedom might lead the child to be unaware or dismissive of errors committed. The right amount of chastisement or praise demands attention and perspicacity.

Lately, the girl has mostly outgrown doll playing. She now has 'activities.' She does gymnastics, is learning piano playing and does bobbin lace, a very demanding traditional craft unique to the region. She, and other children have been portrayed as examples of the revival of an important local craft, anchoring her self-worth and identity in the ex-libris of the region. Nevertheless, the girl loves clothing, and rehearses her wishes and color choices as she dresses her dolls (Sutton-Smith 1986: 194). Recently, the girl received first communion and, when prompted for a choice, she specifically requested a party instead of presents.

The complex interactions played out between children, toys, and traditional festivities and foods demonstrate the cultural interconnectedness between food and toys and the merging of experiences that mark childhood memories whereby foods, and sometimes a toy, become lifelong signifiers to be called on for home and security feelings. Further, a doll's naming ceremony (baptism) becomes a sort of reenactment of the child's own rite of passage, an occasion on which toys and food are central to the celebration. All these events that include gift giving and special food displays are occasions for mothers and children to comment, evaluate, and gossip on income, ingenuity, status, and social competence.

The Economics of Food and Toys

Despite the Portuguese economic woes, the endless cycle of children's birthday parties and other special occasion festivities, with the consequent exchange of presents, cakes, and displays of food, emphasizes the importance of these ceremonies for social status and acceptance as well as the relevance of teaching children the rules of social engagement and networking. These are no longer solitary, homely moments of practice and learning in which mother and child interacted, as they would have been in another century. These are now public performances that carry social weight and can result in sanctions as much as they can produce lifelong friendships. There is a lot at stake.

In times past, children were seen, not heard. If a child was not cute or smart enough, he or she was not much seen either. Children lived in the silent world of mimicry to become miniature but convincing adults. They were offered no choices; they were said to be compliant or reticent, easy or difficult. They remained silent during meals, spoke only when questioned, and asked permission to speak or to leave the dinner table. Generally, in Portugal this view of children did not fade until the late 1980s. Today, they make choices and their active social life engages parents in a rollercoaster of activities. Several parents I talked with expressed their frustration with this trend of birthday parties, its cost in food, and the difficulty of procuring appropriate but reasonably priced gifts. Though these party displays are by children and for children, they are also the mothers' showcases and testing grounds; mothers are assessed at many levels of competency and status. These displays entail a delicate balance between inventiveness and economic power, and require mothers and children to be attentive to the clues of acceptance or rejection.

The competitiveness of these displays demands that children become skillful in dealing with social rebuffs and with the shifting position of social influence they occupy in the group. Mothers, on the other hand, must demonstrate ingenuity if not wealth. A complex exchange network is embedded in these displays. Mothers barter skills and possessions (cake making and decorating, candle making, loaning of large pots and pans and tablecloths, patio furniture, baby-sitting, transportation, etc.) as they maneuver around the economic difficulties the festivities pose. Such borrowings and services are tacitly known among the participants, but are seldom discussed. Need supersedes competition as mothers want to fulfill children's expectations as well as their own.

It is not easy to evade this network of celebrations and gift giving, as reciprocity is embedded in the Portuguese ethos and mothers become ensnared through school and church activities, parents' meetings, and local affairs. As children grow up and the circle widens, so do the demands of reciprocity. But the Portuguese were born into this 'game,' though manifestation of its rules as well as its structural practices and economics have changed with time (Bourdieu 1990: 68, 75). As children, mothers were inducted into the game and heard the hush-hush conversations about what to give, who deserved it and who did not—the politics and affects of gift giving within the extended family. They learned the intricacies of exchange within the family and the gender and age differences embodied in gifts, a 'materialised system of classification [that] inculcates and constantly reinforces the principles of classification which constitute[s] the arbitrariness of a culture' (Bourdieu 1990: 76). Like most subjects in a culture, women want to do better when they too become mothers. They grow up indexing what they will do differently, how they will subvert the game of giving by surpassing it. In consequence, these celebrations are also occasions for mothers to compensate for what they did not have as children or to play out their own memories and fantasies. Overcompensating is historicized and criticized. In Caxinas, the stories hint at fishermen lost at sea, poor fishing seasons, abandonment, excesses in consumption, and living beyond one's means to show off and be accepted. These issues may be even more relevant for, and more felt by, immigrant mothers who will use these occasions and foodmaps to pass on traditions to their children and engage in conversation with the local community of mothers. While immigrants from other countries benefit from being culturally 'other,' Portuguese who move into the area from other regions or who emigrated to other countries and later returned are harshly judged for their lack of cultural knowledge, their language deficiencies, and lack of practical acumen.

An example of the economic dilemmas lower-middle-class families face is the mother of the girl from Caxinas. Earning slightly more than the minimum wage in Portugal and raising two youngsters, the mother considers carefully her purchases and presents, and demands commitment and use of the object be it a 'toy,' a musical instrument, or the accoutrements to learn bobbin lace. Though the child's parents struggle in the Portuguese financial crisis, they would not consider emigration: their skills would probably not transfer well to the European labor market but, even if they did, both parents would consider emigrating a disgrace. In general, emigration is an option of last resort, reflecting economic stress, frustration, and the anguish of becoming other. Emigration for economic reasons entails loss and embarrassment and carries the acute feeling that one was not good enough to 'make it' in the homeland.

Last summer I witnessed several young people who were emigrating to European and African countries because of the work and salary opportunities those countries offered. They were in tears, and all were determined to return. Often they do not, as they organize their lives and establish families elsewhere. If they return, it is often later in life. Boring, but very popular, daily television programs vividly illustrate the impact of past waves of emigration as older family members send messages and hugs during those programs to those who remained abroad—second and third generations. These television programs are seen throughout the Lusophone world and bank on the nostalgia of regional traditions, particularly food, music, and crafts, to attract a local audience of older first-generation emigrants who returned to Portugal, as well as their kin who remain abroad. Displays of abundant regional foods shown in traditional ceramic vessels, and mouth-watering sampling and conversations about ingredients, recipes, smells, tastes, and festivities associated with the foods tug at people's memories and yearnings.

The first generation of immigrants who left home as adolescents, or at a later age, dream of specific homeland foods. Their cravings are felt on the body and the subconscious and go beyond nurturing and identity needs, *and may resemble those of starving people.* Particular dishes may be associated with weekends, calendar dates, or with specific locales and events which occurred seasonally or daily. In the geography of childhood, these contextual spatial and cultural associations with foods create a sensory network that makes memories smell and taste, often making some foods unrivaled. This mythologizing of dishes becomes the core of family traditions. I recall that when I came to the United States I was asked what Christmas traditions my family (my United States family) had. I was disconcerted—I had none. Having been born elsewhere, my 'traditions' were untranslatable into American foods. Description of Portuguese Christmas foods provoked equivalencies that reduced images and tastes—but particularly desires—to trivialities: Portuguese Bolo Rei is not fruitcake, and rabanadas are not French toast. Toys and the preparation and consumption of special holiday foods are implicated in the creation of traditions and the longing for home. Bolo Rei is a Portuguese traditional Christmas and New Year's dessert as are rabanadas, also called golden slices (*fatias douradas*). My refusal to entertain similarities speaks to the identity importance of these foods.

Though homeland and families are conflated with food in quotidian and ceremonial celebrations, mothers have traditionally been at the center of affect and memory as, generally, they are charged with nurturing and food preparation practices. Research undertaken in the Midwest of the United States showed that Latina immigrant mothers were the primary caregivers. These mothers struggled to retain their cultural food traditions while making sure their offspring adapted to the new food environment without becoming addicted to 'fast foods.' Mothers strove to make sure the children ate a healthy diet and did not lose food connections to their homelands (Greder et al. 2012).

First-generation immigrants had traditional foods available in the shops and restaurants in their homeland neighborhoods and could elect to have them at any time. In the host country, immigrants have to procure these foods, or the ingredients to make them, in specialty shops often at prohibitive costs. They can also live in ethnic enclaves where local markets cater to specific populations, or they may not have traditional foods available at all. The lack of regular availability means the second generation has to acquire the knowledge to prepare those foods within the household or progressively disengage from them (Koç and Welsh 2002: 4; Soo 2010). The children of first-generation immigrants have a different relationship to homeland food from that of their parents; food traditions are learned within the household but in the context of a foreign country, and are experienced as treats often to be had on special occasions. Quotidian fare metamorphoses into the exceptional, and the soothing effect of the habitual is lost. Traditional dishes are no longer comfort food to be whipped up when longed for, but become identity symbols marking special occasions, as if identity was a 'feeling' one had to reignite or reaffirm at certain times.

Although both generations display identity in and through the ceremonial preparation and sharing of traditional foods, the second generation of immigrants is less attuned and concerned with the authenticity of ingredients and with taste. Thus, their identity as Portuguese is inscribed through the ceremonial process and the conviviality more than from the outcome. For them, traditional foods are proxies for identity display, but they also embody the separation between affect and effect, between the taste of the habitual and the precariousness of the treat. This transformation of the habitual and taken-for-granted into the exceptional and the contrived parallels the process of change between the first-generation expatriates and their offspring.

The balance between integration in the host country's culture and maintenance of the heritage culture is difficult. As Weinreich (2009) demonstrated, the permutations between situations of acculturation and enculturation to the new country and retention of old country culture are many and complex and will vary with biographical experiences, gender, age, country, region within the country, and certainly with the elements to be enculturated and those heritage traditions to be retained. Another study that considered the adaptation of recent immigrant families from India and Pakistan to Canada showed that the children of first-generation immigrants adopted almost completely Canadian foodways, including traditional Western holidays, while their parents continued to eat and prefer native foods. To foster a sense of identity and community and reinforce traditional ways of behaving, parents organized potluck dinners and religious events. Despite the adoption of Western foods and other customs, the study found that the children of these immigrants had a good grasp of their traditions and values though often disregarding them (Wakil et al. 1981: 937–938). As first generations of immigrants hold onto traditions and foodways to fortify their identity amidst cultural change, their children negotiate the fluidity of their identity as citizens of the host country through participation in the multiculturalism of consumption (Caglar 1997: 182). I have spoken of parents and kin but have avoided notions of nuclear families, as their structure, formulation, and representation in modern Europe varies widely (Fonseca and Ormond 2008: 94–96). Nowadays, it is all too frequent that the mother remains working in Portugal with the children, to provide them with stability, while the father works in another European country for long periods and comes home for the holidays or the family joins him wherever he is working.

For the youngsters, toys are essential to this enculturation process as they mediate the distance between the first generation's experiences (habitus) and the inculcation of skills (praxis) that will be translated into memories of a country in absentia. But toys do not stand alone: they need the active involvement of children to become alive with teaching power and to engender experiences. Toys for food preparation, display, and consumption are interconnected and require many objects, as *realistic* make-believe is expensive. Playing at serving tea and cake requires working with many toys and choreographing many tasks. Mother will serve as the facilitator and guide the preparation. As the mother works with the 'real' cooking objects, she merges the adult world with that of the child. In the liminal space of the imaginary, the immigrant child superimposes the host country on the imagined country, the country whose coordinates are special foods and stories and that comes together in the ceremonial world of cooking traditional foods. She also superimposes the real utensils on her mimetic toys. And she learns. It is hybridity at its clearest, because it is unrecognized, unchallenged, and goes without saying. Later the child will know the mimetic trick that has been played on her and that toys facilitated, as reality and make-believe fuse. As she grows and layers social statuses and roles, the perception and value of her ethnic identity will change and be reevaluated and reformulated many times over, but the experiences of childhood and adolescence will be essential to her selfhood and maybe to her self-worth (Hutchison and Smith 1996: 6).

While the child dwells in hybridity, her first-generation immigrant mother will always be in country-limbo. She will belong to both countries and to neither, and will find herself out-of-place in both: always a foreigner and always a migrant. Migration is a condition that adheres to the psyche and becomes second nature, literally. She will be questioned on her competency in both countries and will often question which expatriation is less painful. She will find that those toys have ensnared herself and her child as they served to materialize concepts of gender, labor, and nationality, and created attachments. Toys eased transition in learning and foodways, but they also served to reaffirm the old country's traditions and values, often at odds with the values of the host country. Mother will think back on her struggle to buy toys for her child to ensure the child knew the proper etiquette to receive her friends and to learn the skills to prepare the dishes from the 'old country': in sum, to be accepted when she returned to the homeland. All these motherly efforts were not just about grafting the socio-cultural traditions through toys and maintaining identity; they were about being prepared to return home.

Conclusions

Material objects do immaterial things; they create affects and memories and change the subject/doer. Toys and foods do that; and playing with toys and food models the self as it establishes, and literally embodies, symbiotic relationships, 'immaterial attachments made within the material world' (Buchli 2016: 1). Food needs toys to turn food into play and sidestep the prosaic act of eating. Engaging children and later adults in food preparation practices enculturates the former and co-opts the latter. Toys are proxies in the nurturing and socialization of children as 'culture is governed by unconscious, embodied and habituated actions,' (Küchler 2005: 211) and after life's subtractions and additions, adults are left with memory remainders. This chapter focuses mostly on the Portuguese, but regardless of where one dwells, toys and food are deeply implicated in the process of becoming, and first-generation adult immigrants bring to the host country defined notions of what an identity cultural profile should be. Immigrant families, and particularly mothers, struggle to maintain the desired level of attachment to homeland food traditions that confer on their children habits that bespeak identity, but their children struggle to assimilate and be different while avoiding being 'other.' These two processes of adaptation are dissonant and create anguish and frustration in families, which add to the inevitable generational cleavage and its conflicts. The solution, if there is one, comes with longing and the need to reinvent traditions and make identity material. In the end, you are what you eat because you eat what you are, or want to be. Or, maybe because you remember what you were.

References

Adams, M. (2007). *Self and social change.* Los Angeles: Sage.

Block, E. (2006). *Traces.* Stanford: Stanford University Press.

Bourdieu, P. (1990). *The logic of practice.* Stanford: Stanford University Press.

Brown, B. (1998). How to do things with things (A toy story). *Critical Inquiry, 24,* 935–964.

Buchli, V. (2016). *An archaeology of the immaterial.* London: Routledge.

Caglar, A. (1997). Hyphenated identities and the limits of 'culture'. In T. Modood & P. Werbner (Eds.), *The politics of multiculturalism in the new Europe* (pp. 169–185). London: Zed Books.

Corolle. (2015). *Advertisement for gilt toys.* http://www.gilt.com/sale/children/hello-dolly-feat-corolle?utm_content=dsr_sale_tile_11&utm_source=gilt&utm_medium=email&utm_campaign=dsr4_12182015_12PM_d. Accessed 19 Dec 2015.

Fonseca, M. L., & Ormand, M. (2008). Defining 'family' and bringing it together: The ins and outs of family reunification in Portugal. In R. Grillo (Ed.), *The family in question, immigrant and ethnic minorities in multicultural Europe* (pp. 89–111). IMISCOE-Amsterdam University Press Series.

Giddens, A. (1991). *Modernity and self-identity: Self and society in the late modern age.* Cambridge: Polity Press.

Greder, K., Slowing, F. R., & Kimberly, D. (2012). Latina immigrant mothers: Negotiating new food environments to preserve food practices and healthy child eating. *Family and Consumer Sciences Research Journal, 41*(2), 145–160.

Hutchinson, J., & Anthony, D. S. (1996). Introduction. In J. Hutchinson & A. D. Smith (Eds.), *Ethnicity.* Oxford: Oxford University Press.

Koç, M., & Welsh, J. (2002). Food, foodways and immigrant experience. Multiculturalism program, Toronto, Department of Canadian Heritage at the Canadian Ethnic Studies Association Conference, November 2001, Halifax, Center for Studies in Food Security, Ryerson University. http://www.ryerson.ca/foodsecurity/. Accessed 27 Dec 2015.

Küchler, S. (2005). Materiality and cognition: The changing face of things. In *Materiality* (pp. 206–230). Durham, NC: Duke University Press.
Miller, D. (Ed.). (2005). Materiality: An introduction. In *Materiality* (pp. 1–50). Durham, NC: Duke University Press.
Neruda, P. (1986). *Winter garden* (Jardín de Invierno) (William O'Daly, Trans.). Port Townsend, WA: Copper Canyon Press.
Ngai, S. (2005). The cuteness of the Avant-Garde. *Critical Inquiry, 3,* 811–847.
Rowlands, M. (2005). A materialist approach to materiality. In D. Miller (Ed.), *Materiality* (pp. 72–87). Durham, NC: Duke University Press.
Russ, S. W. (1993). *Affect & creativity: The role of affect and play in the creative process.* Mahwah, NJ: Lawrence Erlbaum Associates.
Soo, K. (2010). *Newcomers and food insecurity: A critical literature review of immigration and food security.* M.A. thesis, Ryerson University, Toronto.
Sutton-Smith, B. (1986). *Toys as culture.* New York: Gardner Press.
Trend, D. (2013). *Worlding: Identity, media, and imagination in a digital age.* Boulder, CO: Paradigm Publishers.
Wakil, S. P., Siddique, S. M., & Wakil, F. A. (1981). Between two cultures: A study in socialization of children of immigrants. *Journal of Marriage and Family, 34*(4), 929–940.
Weinreich, P. (2009). Enculturation, not 'acculturation': Conceptualising and assessing identity processes in migrant communities. *International Journal of Intercultural Relations, 33,* 124–139.

Author Biography

Mariah Wade is an Associate Professor in the Department of Anthropology at the University of Texas, Austin, USA.

PART IV

Toy Design and Play Spaces

CHAPTER 14

Work and Play in a Theme Park

Luísa Magalhães

Introduction

Theme parks, as replacements for older amusement parks, have recently become a frequent leisure destination for families with children. As well as families looking for opportunities to give their children more and more possibilities for having fun outside the household, this is also a quest for education outside the school environment. Children and adults enjoy the park for a variety of reasons, including the skilled staff who every day create such an enjoyable environment.

Beneath the cheerful environment created for players at the KidZania Lisboa Park there remains a certain amount of ambiguity, related to its ultimate purpose of generating profits. This ambiguity refers to both brand education and free play. Adults often express tension through some expressed criticism and complaints about the marketing strategies in the park.

Observation and research carried out during park activities show that children express major appreciation of being able to perform adult tasks, such as working and earning money with the consequent decision of how to spend it, disregarding the possibility of being early victims of branding,

L. Magalhães (✉)
Catholic University of Portugal, Braga, Portugal
e-mail: luisamagalhaes@braga.ucp.pt

L. Magalhães and J. Goldstein (eds.), *Toys and Communication*,
https://doi.org/10.1057/978-1-137-59136-4_14

of which they are not even aware. By promoting gradual acquaintance with real-life objects and situations, KidZania claims to be promoting a particular educational format for play based upon communicating facts such as market language and cultural endeavors. The park engenders the possibility of creating a sense of productivity at an early age, providing children with the feeling of achievement that comes from their being capable of earning their own money, using it and spending (or saving) it. In this sense, the world as they know it does not exist without brands.

The confrontation between children and the adult world of brands should not be considered as an educational problem for players. Brands are pervasive and ubiquitous in today's developed adult world, engaged in market formulas that are transforming adults' behavior, not only about themselves and their habits of consumption, but also about the way they raise their children. The brand logo is used and explored in order to give children a feeling of fulfillment and happiness that eventually comes out of their gained capacity to purchase. And this is a rewarding capacity for children as well as for adults.

If the perspective of branding were left aside, the KidZania Lisboa Park would resemble a uniquely designed make-believe play toy park. Users/players would come and enjoy a whole world of pretend play, using their favorite toys on an adult scale as tools for entertainment.

There are, however, several brands involved in the designing and functioning of KidZania. These are, in fact, corporate sponsors for the corresponding activities in the park, and players become closely acquainted with their commercial logos and products. These become easy to identify in the outside world as well. Therefore, it is currently accepted that this park involves a strong branding strategy that theoretically aims at creating early consumerism.

This chapter presents research that was performed on site by presenting an inquiry to park users. It was intended to check on the awareness of players towards brand schemes as well as to check on the hypothetical existence of brand pressure during play. In fact, it is assumed that, according to David Buckingham (2011),

> Children are not either *passive* victims *or* empowered and autonomous social actors. Consumption is not simply a matter of manipulation and control or of choice and freedom. Consumers are not simply 'slaves to the brand', but nor are they joyfully creating their own meanings—let alone express resistance to the powers that be. (Buckingham 2011: 226)

The meaning we believe to be created and strongly developed in the KidZania park is *play*. And entertainment is a special goal for the park, even though marketing strategies are not to be underestimated. But the educational benefit is stronger than the educational, since the park explores particular possibilities of training in several jobs that require a strong sense of citizenship, such as *police*, *court of law* and *university lecturing*. Such is the effort to promote *edutainment* in the park: the ideal mix between education and entertainment.

Even though it may be acknowledged that the educational benefits of edutainment can be hard to prove, the same may be said about advertainment and all other terms related to the act of having fun while learning takes place at a certain period of life.

Children remain a very desirable target, not only as present consumers but also, or particularly, as future consumers. They should be educated as such.

About KidZania

KidZania is an international theme park and is presented through several traditional and new media, including social media—e.g. YouTube, Facebook, and their own website.[1] The objective of KidZania park is to help 'change the world' and to promote world peace and social equity. The existence of culturally distinctive characteristics within the different parks around the world, such as language and economic systems and governance, can also be observed as a peculiarity of this theme park and this can lead to the development of curiosity and personal engagement from children as players.

In order to 'change the world' some social responsibility factors are to be considered. This is stated in the KidZania website which explains that several themes are still under discussion in the park facilities.[2] Citizenship, for example, is promoted in areas such as health, environmental sustainability, recycling, road safety, renewable energy, energy efficiency, and innovation.

As a thematic park KidZania provides a universe of different adult daily life and working contexts for children to enjoy in pretend-play structured scenarios. In formal terms, the park consists of a restricted area in which a model city has been built to children's scale and dimensions. It includes practice and execution of adult jobs and it implies the development of diverse skills regarding the fulfillment of adult tasks in different professional environments that are built as a replica of adult real-life working contexts. The Lisboa park provides over 60 different working roles.

On arrival at the park, children must *check in*; they are given a *boarding pass* and a *50 KidZos check*, and they can *work* as if they were adults, choosing from more than 60 different *jobs* in replicas of some of the most representative institutions of the real city. These include, for example, airport facilities, factories, shops, racetracks, police stations, firefighters, media, health facilities, and several other contexts of work. They will cash their check at the nearest *bank*, a replica of a well-known national bank. KidZos are a specific currency designed to be used at this city, in various jobs, both as salary and as a medium of trade. Children are encouraged to spend this currency after they've earned it in the various spots designed for play and leisure that also exist in the city. Therefore, the money they receive will later be used, either in shopping facilities, or anywhere else inside the park, according to the child's will. It will soon run out, which will make kids *apply* for jobs in the attempt to earn more money and participate in several other paid activities in the park. Hence, there begins the *professional* challenge, as kids are given the possibility to *work* according to their abilities and particular preferences among the 60 existing different jobs.

Jobs are adapted to a series of age-gender criteria and foster success on several distinct levels to a wide range of visitors. Boys and girls are free to choose their favorite profession, and are allowed to spend or save their money using their *bank account*. The adult working contexts involve the sponsorship of a variety of brands as well as a series of reproductions of public and private institutions. These explore children's own "labor" skills in order for them to imitate the adult world. This imitation includes a whole range of possibilities regarding several institutions as well as market choices.

The fact that KidZania is a multinational chain of family entertainment parks, like Disneyland or the Coney Island amusement park, cannot be denied. It remains, however, an original characteristic of KidZania that it fosters the make-believe behavior of adults within specific adult environments that have clearly been downsized to kids' levels and degrees of civic understanding. So, activities are simplified and made fun through the possibility of taking different adult roles and making decisions in a somewhat made-up world.

According to Scott Lukas,

> The aim of the park is to create what it calls 'edutainment', or the use of entertainment technologies to teach children about professions in the world. The aim seems very innocent [...] the sinister side is found, however, in the

> powerful branding connections that are established in all of the park's simulations. (Lukas 2012: 176)

In fact, kids are said to take consumerist action at an early age, this being a negative perspective about the park's activities. Apocalyptic views about the generality of theme parks usually point to some market engagement which can lead to kids being induced to play in a corporate-sponsored environment. Although kids may hardly be aware of such inducement, they could ultimately be branded for the future, their future behavior as consumers thereby being conditioned.

According to KidZania's founding father and creator, most theme parks are involved in marketing strategies and this cannot be escaped since brands are also park sponsors. Xavier López created the first KidZania park in Mexico City in 1990, gathering sponsorship from several well-known brands and information from the UN about the average teenage population of each country where he built the parks. However, no other park includes such concern about language learning, road safety codes, the value of higher education, instructional speech, and adult participation in simulated real-life situations, such as firefighting or court trials or even university lectures. In fact, the play activities at KidZania are not simply make-believe and not simply pretend play: they are real branded play activities in which learning is developed in action and this goes together with the possibilities that are provided by the existing brands on site.

Once in the park, kids walk around in the streets of the *city*, searching for indications, since the only thing they know is that they must cash their check in order to start spending their money and enjoying the park activities. After going to the *bank*, queuing and filling in the necessary *forms*, they receive their money and, against its deposit in a *bank account*, they are given their *debit card*, which they can use at the nearest *ATM cash machine*. So, they start spending their money in some park activities, soon to find out that not all activities are to be paid for. Instead, some jobs can actually be performed for payment—and therefore each bank account may receive more *savings*. The only thing that kids must do is ... find a job and get a salary in the park, thereby entering a make-believe adult world in which having *fun is* a major objective.

Toys in the Park

When someone first observes the development of the activities in the KidZania park, the first reaction immediately calls upon marketing effects on defenseless children. This is due to the unavoidable and undeniable presence of several well-known corporate brands distributed all along the city *streets*—these brands being exposed according to their corresponding presence in each country and therefore well identified by children. In fact, both brand content and brand shape are well exposed in the park. Each area *resembles* the authentic area that children know from accompanying their relatives and friends to shopping or playful activities and therefore children are *sensitive* to the acknowledgement of every place in the *city*, i.e. in the park. Things feel credible, real and authentic, as Lindstrom puts it when exploring the concept of sensorial marketing (Lindstrom, 2005). So, the supermarket is a familiar choice for children, as well as mobile networks or even restaurants and pharmacists, which they know from daily life. Marketing effects would then be quite obvious, as children would easily copycat their behavior when they are in daily life outside the park, perhaps even pestering for something they might have done there, playing *as if they'd have bought something*, only to ask parents for its actual purchase.

However, if the first observation is clearly conductive, it may be misleading or even double-sided. Therefore, our research position has been designed in a way that, while acknowledging the obvious marketing communication strategies it also evidences another array of conclusions regarding daily routines, real-life experiences and also traumatic extraordinary events. In fact, dealing with *traffic jams, accidents, shopping, animal care,* or *schooling* are socially relevant behaviors that can be trained—which parents do, to provide children with healthy social growth. This educational purpose implies familiarity with traumatic experiences that demand for quick action, as occurs in the witnessing of accidents, for example. Also, dealing with illness, with the birth of siblings and corresponding parenting difficulties, as well as accepting treatment in hospital environments without panicking, are only some of the necessary skills to develop in adult life.

Moreover, when observing and questioning these children, there occurred simultaneous suggestions for the use of this theme park as a positive instructional area for exercising coping skills and, at the same time, getting *adult satisfaction*—i.e. the satisfaction that children often experience and explain when they are asked to behave in adult roles.

Other less dramatic but also necessary real-life jobs are introduced to children in the park. *Firefighters, policemen, ambulance drivers, window cleaners, street cleaners, court officers* are, to name a few, some of these other jobs that may not be as readily acknowledged by children. These are brought to light at the park as useful and dignifying jobs, on which people can rely, not only because of societal needs but also because of their impact and relevance in everyday life contexts. Valuing these jobs remains as an extra surprise when leaving the park, since, depending on their age, children have different views about them.

Research aimed to observe whether children were able to identify the used toys as both objects of play and working tools, assuming the possibility for children to distinguish between *work* and *play*, and also whether children would hypothetically be immune to strategic market campaigns that would later induce guided consumption. Different but cooperative roles were ascribed to the variety of materials in the park: brand logos, familiar colors and textures, customized vehicles and food providers. When understood as *toys* by children, these materials would need to be seen as dynamic objects, with a given meaning. According to their role in the performed activity, materials in use can convey different meanings—about which the remaining question addresses their characteristics as *objects for play* or *working tools*. The separation between these two concepts and corresponding possibilities remains very difficult, if not impossible to assess. Toys are signifying objects and their semiotic value as signs can be inferred at both levels of usage as tools and as elements of play action. In some cases, both possibilities are considered; hence the semiotic definition of *dynamic object* would apply to this situation because it addresses the articulation of meanings that occur during play performance.

The concept of *dynamic object* of a sign is here taken from the work of Charles Peirce (1834–1914), and it appeals to *reality*, implying the possibility of movement, of change and of life itself. The alternative concept is the *immediate object* and it is presented in distinctive terms:

> We have to distinguish the Immediate Object, which is the Object as the sign itself represents it and whose Being is thus dependent upon the Representation of it in the Sign, from the Dynamical Object which is the Reality which by some means contrives to determine the Sign to its Representation. (CP 4.536)

Sign meanings would change when transformed by manipulation and use, therefore developing into another level, always promoting different interpretations and allowing for different forms of play which would consequently imply different representations according to players' skills. For example, when working for food, at a make-believe well-known hamburger restaurant children perform both as workers and as players. They work, because they produce food and they will eventually eat the results of their work. They play *as if*, because they pretend to be cooks or employees serving customers at the restaurant. They have *fun* because they are learning and producing their own food, just like adults often do in daily life.

As *immediate object*, toys in the park serve as tools for performing the required tasks in each area or branded stand whereas, as *dynamic object*, they are play elements that consequently offer the possibility of make-believe, i.e. of *representation*. Hence toys in the park hold an undeniable semiotic value.

Research presents a proposal for the understanding of the triple role of toys in this particular theme park. It ascribes toys the possibility of generating different meanings according to several distinct factors that will depend on the characteristics of each player. It is based on a semiotic profile that was initially developed in order to clarify the meaning of the concept of sign by Charles Peirce (Magalhães 1998). The scheme provides a representation of the idea of transformation between each possibility of generating play meanings according to different semantic as well as operative values that are endorsed by the KidZania park.

Three major signifying concepts were chosen and applied by players themselves to their activities in the park. At some point they are coincident, in the sense that the function of toys, as well as their position in the moment of play, consists in a semiotic representation of the adult world of work. They are the concepts of *play, work*, and *fun*.

Play is the initial aim of the activity that starts and develops into something objectively productive, from which money can be obtained as payment, as well as working reward. Since it addresses a feature of adult *reality*, it becomes *work*, therefore reaching the following level. It ultimately reaches the aimed goal of *fun* because it gathers the first levels into the wider level that subsumes the park and the results of the visit. This remains a never-ending spiralling circuit, or better, a circuit that only ends with the exit from the park environment (Fig. 14.1).

Peirce, C. S. (1931/1958). The representation of the circulation of meaning through the Sign-Object-Interpretant relationship established by Peirce is hereby mirrored in the relationship between Play-Work-Fun. The image reflects the spiralling effect that can be observed during the type of

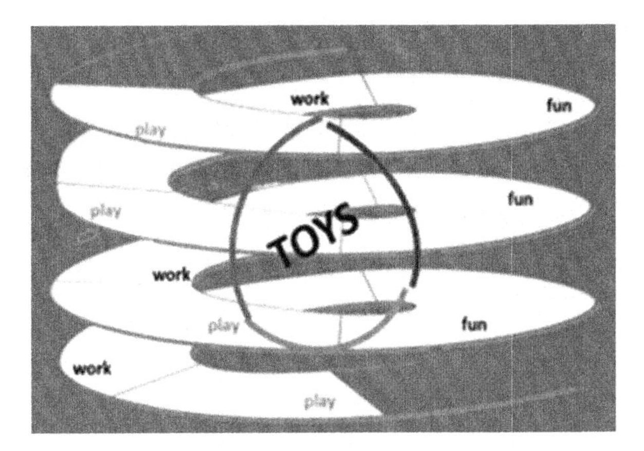

Fig. 14.1 Semiotic representation of the concepts 'Play,' 'Work,' and 'Fun,' inspired by Charles Sanders Peirce's trichotomy of signs

play activities that are promoted at the KidZania park, where child players start playing – pretending to work as adults – having fun when performing *adultlike activities.*

As concepts develop both for children as players and for adults as observers or as providers, it is acknowledged that the first use of toys is of course promoted by the play activities that they suggest. Adults see them as *tools* to keep children busy, to get children learning or to exercise children —as when playing sports games or using building blocks. Different types of play suggest different uses of materials in the park, and different problems in selecting the right ones depending on the immediate purpose. The park is expected to develop children's personal *working* skills while providing some reassurance to adults who wish to provide their children with the appropriate toy to foster adequate development.

On the one hand, children do not care quite so much about the quality of the toy—they can play as well as have fun with other *natural* things from the real world, such as beach pebbles or wooden sticks. Toys are supposed to provide *fun* above all other issues or emotions and to engage children in the pursuit of something truly enjoyable that implies freedom and creativity, regardless of the existing instructions for play, for example. On the other hand, when using toys as productive materials and as working tools, children reinforce the flexibility of some objects that exist in the park and also in real life and that eventually embody the *ambiguity of play.*

Brian Sutton-Smith acknowledged this ambiguity, which he expressed as a behavioral duality, according to which play would foster children's adjustment to their own daily lives (Sutton-Smith 1986). It remains sequential and consequent that the use of toys in this particular theme park implies mastering several skills and developing learning in multiple senses. Adding new uses to old toys or to parts of old toys seems to illustrate the transformative character of toys as play objects, and therefore also illustrates their cultural characteristics Goldstein (1994). To play with toys means to share culture and, in this sense, toys can be considered as elements of cultural transmission. Therefore, cultural values are ascribed to particular types of play.

Roger Caillois (1967) provided a consistent terminology to address play, which can be adapted to the park activities. According to this terminology, *play* can be described within four different perspectives regarding play action: *Âgon* (competition), *Álea* (fortune), *Mimicry* (representation), and *Ilinx* (vertigo). These can be separately and also simultaneously observed, depending on the play contexts. *Play* should therefore be regarded as the most necessary and balanced human activity, combining human instincts of survival with human feelings of joy and achievement. It involves the possibility of enhancing sheer competition (*Âgon*) as well as complex pretend play (*Mimicry*). As long as it is freely conducted, the activity of play lies underneath the sense of fulfillment and allows for the possibility of illustrating the process of growing into adulthood through the sequential steps of *play*, *work*, and *fun*.

Each block is included in the *play* area of the scheme and it regards the *Mimicry* concept developed by Caillois (1967). It consists in performing roles within contexts that are provided by the park. Representation of roles implies the acceptance of the same roles that are ascribed both to players and to materials as well as their sequential use as elements of *play*. In the KidZania park environment, *play* is about *Mimicry*. The active urban world of adult humans is represented through the eyes of the players: it satisfies the definition of the terms of conviviality and interpersonal communication as well as of the terms of education, justice, health, trade, and even kinship. And these are all possible because players act *as if*, thereby enhancing representation/*mimicry*, playing several roles in an artificial context of an adult world.

Research

This research was conducted at KidZania Lisboa Park in order to draw conclusions regarding the role of toys in this particular environment. It led us to question park users in order to understand their actions during play time.

Research was conducted over a weekend, from 10:00 to 19:00, with a group of 300 respondents, age range 5–15 years old. Of these, 77% were 8–11 years old, ranging from third year in primary school to sixth year in preparatory school. Girls outnumbered boys 183 to 117 in the park area.

The research included a six-question protocol, addressing specific matters and corresponding to specific research objectives. Questions addressed items related to present activities and preferences, future choices and favorite toy/tool. Each question was designed to serve different research objectives, as follows:

1. *How did you learn about K?*—Research objective (RO): to identify the source of information about the park.
2. *What do you like to do here?*—RO: confirm expectations and previous knowledge about the job; express reason for job choice.
3. *What are you doing now?*—RO: check priorities and job relevance, according to player's life experiences.
4. *Why did you choose this job?*—RO: assess the knowledge about job characteristics in real life.
5. *What will you do with your KidZos?*—RO: identify purpose for play or for work.
6. *Would you like to come back some day?*—RO: check expectancies about the park.

This set of questions was prepared in order to examine the use of toys within the different *professional* contexts offered by the park. The following underlying questions were triggered by players' engaged attitude: were these gadgets for play? were children really *playing* in the KidZania theme park?

Could it be that children engage themselves in *working as adults* from an early age? Do they think they get involved in *role play* activities or are they really committed to performing a *job*? Are children aware of their activity of play in this park? And what creates the difference between play and work, as long as children consider this *fun*?

Children's preferences about the chosen jobs were registered, and the whole set of questions was presented to them while they were entering new areas of their choice. No guidance was offered as to which *work area* should be chosen, and the content of dialogues and of play roles was made freely.

The individual interviews were conducted according to the script and to each questionnaire in order to test the hypotheses that follow. The answer to the first question tagged the origin of children's knowledge about the park as being mainly approached as word-of-mouth from friends and family. Other hypotheses were as follows:

(a) KidZania toys are seen by child players as *replicas* of both tools and institutions from daily routine contexts that they are familiar with (questions 2 and 4).
(b) By using these replicas in contexts that imitate adult life, players assume an adult role and act *as if* they were adults (question 3).
(c) Interaction occurs within both *player-to-player* and *adult-to-player-to-adult* environments, in which there is no formal distinction between the addresser and addressees—children or adults (question 5).
(d) KidZania objects play a double role that is based on the possibility of being educational and changing a part of the world and simultaneously having fun (question 6).

As children became *players* when entering the park, they were approaching different areas and different types of activities, according to their expressed goals of *having fun* or *making money*. They became *respondents* for a few minutes along their play journey, when they were presented with the above-mentioned set of questions to which they willingly responded.

It has been acknowledged that KidZania theme park is advertised in different media, particularly online[3] and also on TV. The efficiency of media broadcasting is enhanced by the possibility of different available media that address different publics. When questioned about their knowledge of the park and their particular source of information, respondents were given six different possibilities, and their answer was very clear: family, friends, and school were clearly the source and the reason for their attending the park.

Even though there are four different TV spots about KidZania on TV channels available in Portugal, such as Panda or Disney Channel, respondents also explained that their knowledge of the park came mainly from friends (43%). Families create situations of conviviality between youngsters, such as birthday parties organized in the park, which later on may lead to former guests having their own party afterwards, therefore extending the success of their visit. Adults seem to be more directly informed through television and advertising which, in this particular case, provides upgraded information and suggests possibilities of extra educational and occupational opportunities for weekends and holidays.

Engagement through friends or classmates was proved more efficient among the inquired groups. Communicative contexts such as school and family remained the most common to share information about park activities; each working area provided multiple play experiences by involving staff members in the role of instructors and guides.

From the perspective of interactive experiences of play considered as communication occurrences, the whole condition of adulthood offered by the KidZania theme park remains an opportunity to share personal abilities and experiences related to areas of adult life that children are very keen to apprehend. It is the case of attempting to work in the area of public services, which includes medical services, post office, court of law, security services, among others.

Respondents were questioned about their job preferences, as well as about their preferences on where to spend their money. After cashing their initial check and receiving a bank debit card, they answered that they would most likely spend all the money before checking about possible jobs. They learnt that they could earn some more money if they could find themselves a paid job that they should perform according to their capacities and preferences. Apparently, kids/players understand very quickly how ingenious and hardworking they must be in order to earn their own extra money. They sometimes do it in an easier way, e.g. by doing easier jobs, and sometimes according to their best skills and working preferences. Defining a favorite job came as a result of the acknowledgement of the *need to work to earn money* and not as likely as a particular motivation.

The choice of jobs in the park expresses a well-rooted desire to become an adult and, more importantly, to become a successful adult: someone who is capable of performing a job and of earning their own money and therefore ensuring the survival of their family. Gilles Brougère (2004) accounts for this desire and refers to the relationship that is established between young girls and their Barbie dolls, for example, as a relationship that expresses a strong desire to grow and to become an accomplished adult similar to the adult image that Barbie dolls provide. Such a bond finds its correspondence in the park when girls decide to take *model courses* at *fashion school*, and they later go and perform on the *catwalk*, posing for *photographers* and *magazine editors*.

When choosing a job and performing it afterwards, child players are expressing their main concerns and preferences about their future as adults. Some of these concerns are based upon a strong wish to grow as a reliable adult, whereas others are simply based on the desire to replicate their familiar status in terms of group harmony and companionship.

In the attempt to match the expressed desires with the present situation at the moment of inquiry, the next question aimed at establishing the present situation of respondents. The majority of respondents (53%) confirmed that their favorite activity was to *work and have a job*, which they apparently succeeded in doing when choosing their respective jobs. They would respond to other questions according to their preferred activity, and answers were mostly related to the longing to be adult and have an adult job, therefore confirming initial data.

Some materials demand skilled handling, which is why the park provides *trainers* who offer the necessary instructions. This applies to the job that is to be performed and also to the expected attitude. For example, *university lectures* are delivered and tested in order for the player to be given a *diploma*. This diploma will create a difference in certain jobs and wages, generating the idea that academic training compensates the effort in economic terms.

Data was gathered regarding present jobs and activities according to the situation of players at the moment of inquiry. The general designation of park areas was provided by park managers and was adopted for questioning: *services* include several possibilities grouped by the park under this designation. It refers to firefighters, law, medical facilities, teaching facilities, or mail distribution. *Retail* includes supermarkets, shops in general, and small commerce which in some cases is related to traditional shops. *Culture* includes drawing school, bookshops, library, theatre shows. *Media* is the general designation for mediated contact with public audiences, as in television, radio, or newspapers.

In every area, if players study and reach a higher level of knowledge, their future income in the jobs at the park may increase when performing a new job. *To acquire a university certificate may change one's life* seems to be the underlying message. Respondents who chose to acquire a university certificate were more likely to increase the amount in their bank accounts and this produced *wealth* and the possibility of more fun in other shopping areas, including the souvenir shop.

The management of a bank account turned out to be a most challenging issue: the use of currency induces a whole decision-making process that may be implied by taking advantage of a given facility—e.g. the *football stadium* or the *racing car track* mostly chosen by boys, or the *hairdresser* and *beauty spa* or the *fashion show* mostly chosen by girls.

Having *money* in a personal *bank account* also induces more imitation of adult life models according to the educational level and family values. For example, when the main objective becomes to save money for the future, using a bank account or cashing the earned salary and considering this as the *correct* action to take in spite of all the temptations and curiosities in which money can easily be spent. However, it can also be credited to family and education when the major goal becomes to go shopping and use the money in alternative options, such as going to the theatre or having fun at a disco bar—the *random purchases* designation includes buying tickets or disco entrances and drinks.

In either possibility, the experience of following the park rules is a very meaningful one. On the one hand, players have *fun* because they endure adult-like jobs built to their own scale and promoting learning and skill development in several different areas. On the other hand, children and adults share the concern and the need for a well-structured market, from a very early age. Learning how to distinguish the brands and their aptitudes may foster players' development of consumer literacy and remains a desirable skill in the adult world, since it is not possible to escape consumption needs. Finally, addressing market pressure from a communication perspective is also an advantage for young citizens in approaching their role in the active world.

Conclusion

In the attempt to re-check the previously stated working hypothesis, the conclusions point out toys in the KidZania park as replicas of existing facilities and well-known brands. This indicates the possibility of recognition of both visual and functional characteristics that enable players to enter into their roles as park users. For example, when shopping at *Continente* supermarket, a well-known national brand, players, in the role of consumers, know that they need to push a shopping cart to carry their items to the cashiers. On arrival at the cashiers' area, they know they must queue

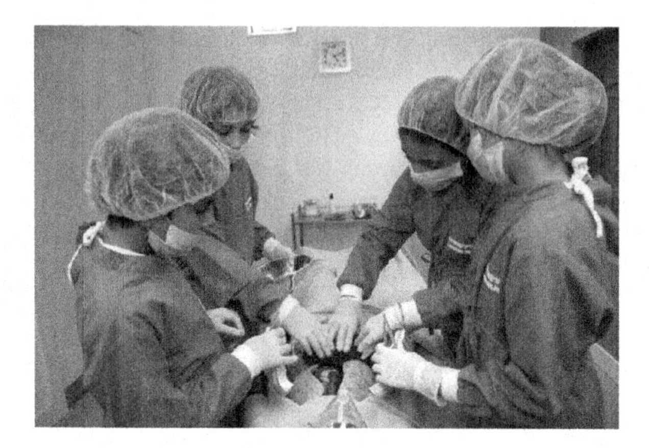

Fig. 14.2 Children at play in the Kidzania Hospital OR, photograph by the author

and wait for their turn. Players also know how to use the *Continente discount card*, which is also a replica of the existing one at the real supermarket outside the park. Also, when at the *Hospital OR (operation room)* players are expected to wear surgical robes and to use sterilized materials in order to perform their job, as seen in Fig. 14.2.

Players in the age range of 10–16 (and maybe some even younger) are aware of brands in the park but, more importantly, they are aware of the adult meaning of each proposed activity. They understand the park facilities and upgrade their knowledge about the working instruments used in several areas and in several different types of job. When combining the age data with data related to favorite activities, it became explicit that younger players (5–8 years old) would encounter their experience in the park as particularly playful and would gain extra awareness of brands when outside with adults. For younger players, it was all about play and imitation of adult endeavors. Older players expressed their views about the adult world in a very direct way: *I thought it was more difficult. I can't understand all adults complaining about work: it's so fun and you can even earn money!* (comment from a 12-year-old girl when leaving the park with her mother). So, it was about work and about performing as an adult all along.

Toys are here recognized as tools for growing and to check on the different values that each job can provide to players. When the job is completely unknown to players, there remains the possibility of being paid and of receiving *adult treatment* by staff. This makes them assume different characters and become *adults* because, after all, they are managing to survive in the park regime as if it were an isolated place with no contact from families or adult caregivers.

When entering the park, children *become* players and they *act* and *learn* according to their personal capacities and disabilities. In other words, they enter a specific narrative in which they are the main characters and they sustain all the events.

The need for *instructors* in some park areas confirms the pedagogical concern that may once have led to the creation of KidZania. Under the role of *instructors*, adults replace teachers but they still *teach* and interact in order to develop players' knowledge and communication skills.

Research found a strong willingness to return to the park and even some cases of recurrent park attendance (12%). Respondents were very direct when expressing their will to revisit, on the basis of not having finished their experience in every area or even of not having performed well in some areas. So, the need to become an adult and the possibility to act like one are the more reliable reasons, behind which lies the possibility of *having fun*.

Brands remain separate from the decision to come back to the park. Some brands were acknowledged and followed, but some others were not acknowledged and were followed as well. Despite brand pressure—for example, in media and supermarket areas—one can hardly argue that this pressure had effective results on child players. Further research would include follow-up contingencies in real market contexts.

As far as communicative interaction is concerned, the observation of groups of players organized in reference areas, such as the *police station*, the *court of law*, the *university*, or the *driving school*, evidenced a deep feeling of trust and solidarity. Older players would help younger ones, providing advice and lecturing them on how to proceed. In some cases, adults in the park would also participate in activities in order to help their children, although this is not a fostered attitude inside the park. This attitude from some adults expresses the *playfulness* of the site and the difficulty to escape leisure moments.

On balance, when in the park, there is a whole narrative that is conducted towards the meaning of adult life. Kids become specific characters in this narrative, by recreating the adult world of basic needs and enjoying the fact that they are learning how to make general market choices. Narratives that are built in every section of the park refer to spaces and action opportunities staged according to the expressed choice of the player who is then invested with his/her character, in order to move on as an *adult*.

Notes

1. http://www.kidzania.com/ (Accessed: 2016).
   https://www.facebook.com/KidZaniaLisboa/?fref=ts (Accessed: 2016).
   Carla Monteiro (2015) KidZania em Lisboa. Available at: https://www.youtube.com/watch?v=hi6K-xCntjc (Accessed: 2016).
2. http://kidzania.com/corporate-philosophy.php (Accessed: 2016).
3. Expresso Online (2007) *Lisboa recebe cidade a brincar*. Available at: https://www.youtube.com/watch?v=S8A43fioE6M (Accessed: 05/08/2016).
   Mafaldaricca (2008) *KidZania Lisboa. Available at:*
   https://www.youtube.com/watch?v=EDfPPCYhQu8 (Accessed: 05/08/2016).

Acknowledgements Luísa Lima, Paula Macedo, Mariana Fernandes, for helping with the inquiry. Marta Amorim Candeias for help with exceptional guidance and access to the park.

References

Brougère, G. (2004). *Brinquedos e companhia*. Sao Paulo: Cortez.
Buckingham, D. (2011). *The material child. Growing up in consumer culture*. Cambridge: Polity Press.
Caillois, R. (1967). *Les jeux et les hommes,* Folio essais. Paris: Gallimard.
Goldstein, J. (Ed.). (1994). *Toys, play and child development*. Cambridge: Cambridge University Press.
Lindstrom, M. (2005). *Brand sense*. London: Kogan Page.
Lukas, S. (2012). *Theme park*. London: Reaktion Books.
Magalhães, L. (1998). *De signo em signo, O signo icónico na obra de Charles Peirce*. Unpublished M.Phil. dissertation, University of Minho, Braga.
Peirce, C. S. (1931/1958). *Collected papers* (Vols. 1–6). In C. Hartshorne & P. Weiss (Eds.). Cambridge, MA: Harvard University Press.
Sutton-Smith, B. (1986). *Toys as culture*. New York: Gardner Press.

Author Biography

Luísa Magalhães (Ph.D.) is a researcher in Communication Sciences, CECS—Communication and Society Research Center, University of Minho, Portugal, ITRA (International Toy Research Association) board member and Congress Chair at the 7th ITRA World Congress, Toys as Language and Communication (2014). Luísa also works for CEFH – Center for Philosophical and Humanistic Studies, Catholic University of Portugal since 2011.

CHAPTER 15

Design for Rebellious Play

Lieselotte van Leeuwen and Mathieu Gielen

Introduction

Saying 'no' was described by René Spitz in 1957 as a spectacular intellectual and semantic achievement in early childhood. The positive description of non-compliance as emerging verbal self-assertion and autonomy is rather an exception in developmental psychology. Most psychological research addresses non-compliance in its dysfunctional forms which hinder social and cognitive development rather than support it (Grusec and Kuczynski 1997; Sonnentag and Barnett 2013). Likewise, the vast majority of toys and games provide information regarding their intended use. From building instructions to game rules, there is a clear set of expectations children should conform to in their play with the respective toys. However, in the realm of play, the joy of breaking rules, defying expectations and worshiping rebel characters is well known (Caillois 1958; Gordon 2008; Salen and Zimmerman 2004; Stevens 2007; Sutton-Smith

L. van Leeuwen (✉)
Department of Psychology, University of Sunderland, Sunderland, UK
e-mail: lieselotte.van.leeuwen@gu.se

M. Gielen
Department of Industrial Design, Delft University of Technology, Delft, The Netherlands
e-mail: M.A.Gielen@tudelft.nl

L. Magalhães and J. Goldstein (eds.), *Toys and Communication*,
https://doi.org/10.1057/978-1-137-59136-4_15

2009). With the exception of children's stories which regularly and famously embrace rebels, there are few play objects explicitly supporting or exploring non-conformist behavior. In the following, we employ two psychological research perspectives, that of development and that of motivation, in order to explore whether there is an argument to be made for rebellious play and, if so, how design for play can contribute to this.

The WHY of Rebelling?

In everyday life, children observe and perform acts of both adherence to and the breaking of rules. Turiel (2003) points out how resistance and subversion are common, due to the level of inequalities inherent in social relationships, such as those between children and adults, peers of different age, gender, and socio-economic status. Piaget (1997/1932) was the first to note that rules are not passively learned but constructed in complex social interactions. Everyday interactions with parents, siblings, peers, and teachers in culturally and sub-culturally determined contexts influence the process of learning to accept or resist specific rules in specific situations.

Increasingly, it is recognized that the ability to resist social pressure is part of a healthy action repertoire (e.g. Schaffer 2006). Automatic compliance does not indicate healthy development or social competency. For example, conforming to a bully in school will not be seen as healthy behavior. Concepts like autonomy, resilience, assertiveness, self-efficacy, self-regulation, internalization of moral rules, and creativity all implicitly require non-compliance to some kind of social demand or normality. Everyday life confronts children with dilemmas which often mean to comply with one party and not with another; to keep playing or comply with adult demands; to comply with being helped or assert one's own competency; to choose loyalty to peers, adults, or to oneself.

Skillful non-compliance, i.e. using constructive strategies to express non-compliance, is a competency related to age-appropriate levels of autonomy (Kuczynski and Kochanska 1990). It takes courage and skill to resist social pressure and to decide when and how to do so. Depending on age and situation, some ways of non-compliance are more constructive and therefore more likely to succeed than others. Play can provide the opportunity to try out and explore the consequences of both compliance and non-compliance in non-trivial, yet protected contexts—the play frame (Bateson 1972; Goffman 1986). Within a play frame, power positions are often redefined; children can assert more power with respect to diversity and type of decisions to agree or disagree with. Play is a context in which children can create rules and openly subvert them. At the same time, strategies of resistance can be discriminated and their effectiveness in different situations tested.

The Developmental Argument

In developmental psychology, the internalization of social and cultural rules to which one should comply have taken center stage in research since the functioning of social groups ranging from families to societies depends on a shared set of rules (see, e.g. Grusec and Kuczynski 1997). Intuitively, there exists a strong association between compliance and morally 'right' behavior and between non-compliance and morally 'wrong' behavior. There are several arguments against such a view: the first being that the application of internalized rules in diverse social situations logically requires non-compliance to social demands which are in conflict with these rules. Secondly, social groups, cultures, and societies are functioning and developing through a dynamic interplay of compliance/agreement and non-compliance/disagreement over a multitude of social interactions (Turiel and Wainryb 2000). Third, in child development, non-compliance, which is typically ascribed to toddlers and teenagers, has been recognized as a healthy sign of asserting autonomy (Kuczynski and Kochanska 1990). Therefore, rather than treating non-compliance as something to prevent, it seems crucial to support children in

- gaining strategies to assert autonomy in socially acceptable ways
- developing skills to evaluate rules and their consequences for social settings
- proposing alternative rules to those that are rejected which respect and integrate the needs of self and others.

The Motivational Argument

What motivates individuals to engage in rebellious behavior? Could there be a world so perfect that people would simply feel no need to rebel? Michael Apter, studying the psychology of motivation, would clearly oppose this idea. In his Reversal Theory (Apter 1982, 2007a; van Leeuwen et al. 2012; Gielen and van Leeuwen 2013), rebelliousness is part of one of

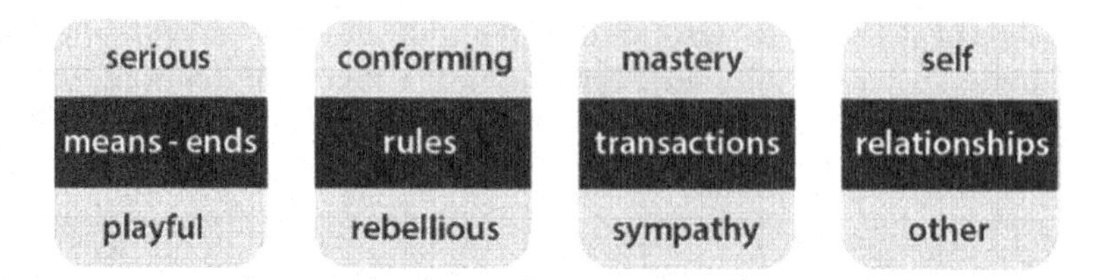

Fig. 15.1 Four dimensions of everyday motivation. Adapted from Kerr and Apter (2007a)

four basic dimensions of human motivation (see Fig. 15.1 for an overview). Reversal Theory assumes that all eight motivational states described need to be experienced and acted on in order to maintain psychological well-being. As shown in Fig. 15.1, the four motivational dimensions each contain two alternative motivational states. At any point in time, we are in one of the alternative states per dimension. The rebellious state, in which rules are seen as restricting and the willingness to embrace freedom and change is high, is the alternative to the conforming state, in which the need to belong, a sense of duty, and compliance to given values are dominant. Rather than being continuously in one state, switches or reversals between each pair of alternative states are frequent in everyday life. Rebelliousness according to this concept is defined as 'wanting, or feeling compelled, to do something contrary to that required by some external agency' (Apter 1982: 198). Both the urges to conform and to rebel can be seen as ways of dealing with social influence (Apter and Smith 1976). This definition leaves the good/bad dichotomy ascribed to both states behind and allows for the broadening of the range of situations in which conforming and rebelling might occur. In line with Apter, rebelliousness can also be seen as social risk taking. From a motivational viewpoint, an individual's intention or urge to rebel is crucial. Not every action that is perceived as oppositional by others has been intended thus. Complementarily, what can seem to an observer like rebellious behavior, e.g. a play fight, is not necessarily the result of a rebellious state of mind but conforms to an agreement between players (Johnson et al. 2001).

Apter views motivation as a dynamic process including sudden changes—reversals—from one state to its opposite. An example would be a child who is playing according to the rules reversing to a rebellious state and wanting to negate the rules when he or she discovers that another player has gained a large advantage in the game which is making other players lose. Actively rebelling against the rules of the game could involve proposing to change the rules, which could turn the luck back in the direction of the other players. A more passive form of rebelling against the other players would be to stop playing and comment that the game is boring before it is finished. The socially expected reaction of accepting turns in the game would very likely still mean that the player is feeling oppositional towards that outcome, but inhibits his or her urge to act on it. The example demonstrates that a negativistic state of mind does not need to lead to negative behavior. As shown in the first reaction, it can lead to creative interventions (see also Gielen and van Leeuwen 2013; Griffin and McDermott 1998; Toscano 2008). Reversals like the one in this example happen frequently and are a sign of healthy functioning. Apter (2007a) identifies the following reasons for reversals: contingency, satiation, and frustration. *Contingency* refers to environmental events such as requests or opportunities for action, and internal events such as memories or boredom, which can induce reversals between states. Examples for inducing a non-conformist state are situations in which children are prevented from playing in the presence of, for them, clear affordances for play. For example, passenger waiting areas at airports often invite running, and train carriages invite climbing. In these situations, parents request children to stay close and be quiet. Switches from conformist to non-conformist states are likely through the contrast between social expectations and physical invitations. Design interventions can detect those situations and enable a playful acting out of the urge within safe boundaries.

Satiation refers to the duration of being in one motivational state. If we have been too long in one state of mind, we might feel the urge to switch without any other reason. A well-known example is the motivation of children to conform to requests of teachers or parents to sit still and concentrate on given tasks. This motivation can only be upheld for a certain amount of time. When required for too long, reversal to the non-conformist state will be induced and sooner or later be acted on. Particularly in the public domain, it is to a degree possible to predict situations of satiation with one state. Children coming out of situations where conformity is required, such as schools, will be more likely to be in a non-conformist state than children coming from the beach. Satiation implies that, to a degree, the need for rebellious play can be predicted from situations and everyday routines. Designs which resonate with a non-conformist state might therefore be most successful in places where conformist satiation is likely, such as near school buildings. For example, a climbable lamp-post wouldn't deny a negativistic state but instead invite its playful expression.

Frustration may lead to reversals in cases where attempts to act according to the rules have been frustrated, as in the example above regarding the danger of losing a game.

Apter's dynamic view of motivation as frequently changing allows us to see urges to not conform or comply as part of the fabric of everyday life. For design, this opens up opportunities of idea development beyond 'right' and 'wrong,' to understand rebellion in relation to other motivational dimensions, and to use play as an opportunity to act out and explore those urges. For a more detailed account, see van Leeuwen and Gielen (2012).

The WHAT of Rebellion

The question of *what* it is that someone rebels against is different from the question about the way in which that rebellion is expressed—i.e. the *how* of rebellion. Design for play can focus on either or both of these aspects. In the following, the two aspects are addressed one by one.

The Developmental Approach

Toddlers' non-compliance, when intended by the child, can be seen as a way to explore one's own action capacities, essentially exploring the social limits within which decisions can be taken (Mireault et al. 2008; Fontana 1988). Consequences of non-compliance inform about opportunities and limits of social agency, just as falling on an icy surface informs about biomechanical limits to physical agency. Picture books like the Alfie Atkins series by Swedish author Gunilla Bergström and 'Papa ne veut pas' (Daddy doesn't want) by Alain Le Saux address this quest for the limits of social action in the form of active negotiation. Translating this into design for toddler play, allowing the testing of each other's limits and providing negotiation spaces for compliance and non-compliance of all sides, are topics worth pursuing in a design for play context.

Turiel's (1983, 2006) 'Social Domain Theory' shows that children from early on differentiate between rules belonging to different types of social interactions and contexts—the moral, conventional, and personal domains. Accepting or breaking rules in different domains has been shown to be associated with different, independent social learning processes (Helwig and Turiel 2011; Smetana 2006). The moral domain refers to absolute rules concerning harm, trust, justice (comparative treatment and fair distribution), and rights. These rules are not negotiable and are independent of situations, institutions, or conventions. An example is the rule not to hurt others. In a play context, moral rules hold as in other contexts. Not acting in accordance with those rules means that the state of play cannot be upheld, at least for those who experience hurt or mistrust or feel that their rights as a person are violated. Examples are hurting someone in a play fight, cheating in a game which does not include cheating (Kushner 2007), or forcing others to do things they do not want to do.

Conventional rules are context specific, based on consensus on what is seen and expected as appropriate behavior in specific situations. Groups like families, peer groups, and institutions regulate their interactions using these rules in defined contexts. When children play, they create often complex sets of conventional rules for the respective play frame from counting-out rhymes to rules which regulate what is being played, how, and by whom. Rules regulate the borders in which play may unfold, such as that each will be in his/her assigned role until they either finish playing or renegotiate. As a group, children negotiate the specific terms of already accepted conventional game rules, such as using the apple tree or the fence as the base to free oneself in a hide-and-seek game. The creation of new conventional rules and subsequent ways to break them are part of nearly all forms of interactive free play (Sutton-Smith et al. 2012). Both the convention by which children organize their play and the rules of games fall within the domain of conventional rules since they are specific for a certain play context. Social play, such as role play, game play, or co-construction play, requires conventional rules in order to function. Those rules can either be given, as is the case for most board games, passed on from generation to generation as in 'hide and seek,' or created in free play. Because of the group and situational specificity, this domain of rules lends itself to exercises of subversion, to exploration of the consequences of making and breaking rules, and to discovering strategies to decide about and/or persuade others to follow or resist rules. Design for play can support a critical approach to rules by limiting the given/fixed rules of interaction and instead provide conditions under which

- agreements need to be reached in order to play
- individual action possibilities and/or states in the game motivate breaking given conventional rules

- the making of rules is the topic of the game, and new rules require the breaking of existing ones
- conventional given rules create dilemmas which bring players into a situation in which complying to the request of one player or group means non-compliance with another
- in a system of rules, players have to identify those which are an obstacle to reaching a goal.

The third domain covers psychological rules, which concern personal issues such as preferences, choices, and self-other borders, but also prudential issues regarding dealing with one's own body. These rules can be broken by the individual him/herself or by trespassing the self-other border of another person. Adolescence is characterized by increased negotiations about autonomy in decision making and rules within families. Not only the following or breaking of rules, but who has the right to set what rules, are debated. Psychological rules are also tightly related to self and identity formation since they provide a framework for the consistency of a person. Taking on roles also means to change the set of conventional and psychological rules by which to act. Designing play contexts in which different (given or created) sets of explicit psychological rules determine action possibilities could provide interesting perspectives on self and others.

Adolescents' heightened negativism/rebelliousness has been associated with the instability of the transition to adulthood and identity formation (Erikson 1968; Cote and Levine 2015; Waterman 1982). Game design has embraced the topic of possible identities and research has shown how, for example, games like 'The Sims' and Massive Multiplayer Online games (MMOs) like EverQuest II allow adolescents to explore possible identities (Williams et al. 2011; Lee and Hoadley 2007; Turkle 2005). An aspect which deserves more attention within design for play is the co-existence of conformities and non-conformities in different social domains. Rebelling against parents, e.g. regarding dressing style, often goes hand in hand with conforming to peer-group pressure. Therefore, rather than addressing rebelliousness in isolation, it seems important to address the dialectics of conforming and rebelling and the resulting dilemmas which characterize identity formation in this age group (Smetana 2006). Design for play could address dilemmas of this kind in a wide range of topics depending on age, culture and historical situation. The topics can be part of the design or chosen by players. Designing for play as rebellion will require working with children in their everyday contexts in order to identify relevant topics or types of dilemmas. We expect contextual design for play to become increasingly important.

The Motivational Approach

According to Reversal Theory (Apter 2007a), the way in which negativistic urges are acted out or not depends on the individual's current motivational state and on the situation and action possibilities at hand. Using the four motivational dimensions in Fig. 15.1, it is possible to identify a number of tendencies. In a playful state, negativism can increase the feeling of excitement, so rules might be broken for fun. If, in addition, in the dimension of transactions an individual is motivated by mastery, then an attempt to cheat in a playful competition could be a way to enact the negativistic urge out, without causing any harm. If in a sympathy state, one might try to increase excitement by convincing others to join in breaking a conventional rule as, for example, in organizing a flash mob or playing a joke on others. Using the dimension of relationships, in a self-motivated mode a person might try to do something they thought they would never do (e.g. dress outrageously)—this means breaking a psychological rule. In the other-motivated state, it could mean joining a rollercoaster ride with a group of friends who normally would not do it. Design could be guided by potential motivational states as demonstrated by van Leeuwen and Gielen (2012). Embracing the rebellious urge as a gateway to break with and return to normality (Lefebvre 1996), to re-interpret contexts, is a motivation which needs to be fostered in each generation. Designing play using potential motivational states as inspiration can provide a scaffold for idea creation and help to imagine playful scenarios in which to challenge Turiel's (2006) domains of conventional or psychological rules.

The HOW of Rebellion

The Developmental Approach

Kuczynski and Kochanska (1990) studied resistance and agreement to parental requests depending on parenting styles. The qualification of rebellious behavior as problematic was related to its frequency but also to the way in which rebelliousness was asserted. In typical development, strategies used to communicate non-compliance indicate increasing autonomy and social competency. From age 4, children will try to reverse or modify their parents' requests and do so in ways which are more or less acceptable to their parents. A 4-year-old who tries to stay up longer by proposing to set the breakfast table for the next morning will have a greater chance to succeed in his/her resistance to the request to go to bed than a child resisting with a tantrum. The emerging ability to effectively negotiate was shown to be related to self-regulation and parenting styles embracing negotiation.

The authors found a transition from more passive to active communication of resistance between 2 and 3 years. Passive non-compliance entails strategies like ignoring or not responding to requests. Active forms of non-compliance are characterized by deliberate resistance to control. Increasing social competency results in a shift in active forms from direct defiance (often accompanied by anger), to simple verbal refusal, to negotiation. Growing up in social environments which practice empathy, perspective taking and respect, support the development of active and more considerate forms of non-compliance.

The ability to actively influence others is as much a part of emerging autonomy as is the widening of physical skills. Group play environments provide many occasions to practice strategies for non-compliance since children's intentions clash at times. Complex environments for free play in particular support the development of negotiation skills. Broadhead and Burt (2012) provide a unique micro-level analysis of the self-guided forms of play and learning emerging from loose parts and materials in a primary school playground in the city of York, UK. The high ambiguity of this context means that children need to create most of the constraints of their respective play frames themselves. Agreeing on a set of conventional rules is part of shared free play. Through the high intrinsic motivation to engage with this environment, children are also motivated to negotiate their intentions in the light of others' requests. The lived autonomy in those 'loose parts' environments (Nicholson 1973; Maxwell et al. 2008) also provides a laboratory for the development of social strategies, including those for non-conformism. Adventure playgrounds are examples of those environments (Norman 2003). But there are also specifically designed parts like 'Outlast' by Community playthings.

In a more explicit way, rules of games are an obvious way to challenge non-conformism either by their pure existence or by rule design. The existence of game rules is in a way an incentive to break them, motivated either by attempts to cheat or by exploring and extending the scope of the game when mastery within the given rule set has been achieved (Salen and Zimmerman 2004). While the first motivation is destructive towards other players, the second leads to new insights.

As a design strategy, so-called traitor games assign players a 'good,' rule-abiding, or a 'bad,' rule-breaking role. Players are not initially aware of each other's role and need to infer from behavior who the traitors are. The same holds for bluffing games which request players to lie. Skillful lying on the basis of others' expectations does indeed exercise strategies of covert rebellion. However, as in the example of the climbable lamp-post, the decision and the basic strategy for non-compliance are taken over by the rules of the game or the affordances of the design. Active strategies of rebelling, like threatening, walking away, using humour, or negotiating, as part of a player's action capabilities in a game could focus attention on decisions to comply or not and on the effectiveness of different strategies to subvert others' intentions.

Students designing for children's play at Delft University of Technology explored strategies for rebelliousness with children and applied their findings to design for play. Figure 15.2 describes strategies for rebellious behavior described by a group of 9–11-year-olds in the Netherlands (Luijpers 2013). Three main strategies emerged in the context of diverse social experiences: sticking to rules but breaking them secretly; pretending to stick to the rules while breaking them; openly breaking rules. Children related the three strategies to different personal circumstances and characteristics that they expressed through quotes such as:

(for the shy little girl)

'I don't like being the oldest. I always have to set a good example and I'm always the one who's punished when we have fun.'

'How am I going to keep this a secret?'

'We like to go to a playground in a different neighbourhood, because our parents are not there.'

(for the popular boy)

'I am very popular, when I arrive somewhere, everyone says "Hey, you're Tom's little brother."'

'Sam always makes fun of everyone. He tries to be the funny guy.'

(for the rebel boy)

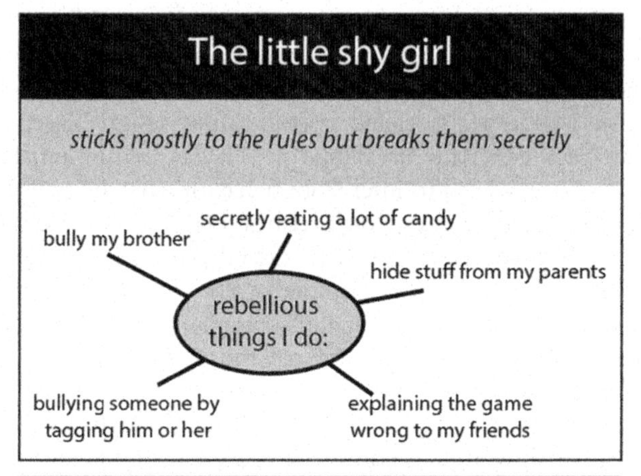
The little shy girl
sticks mostly to the rules but breaks them secretly
rebellious things I do:
secretly eating a lot of candy
bully my brother
hide stuff from my parents
bullying someone by tagging him or her
explaining the game wrong to my friends

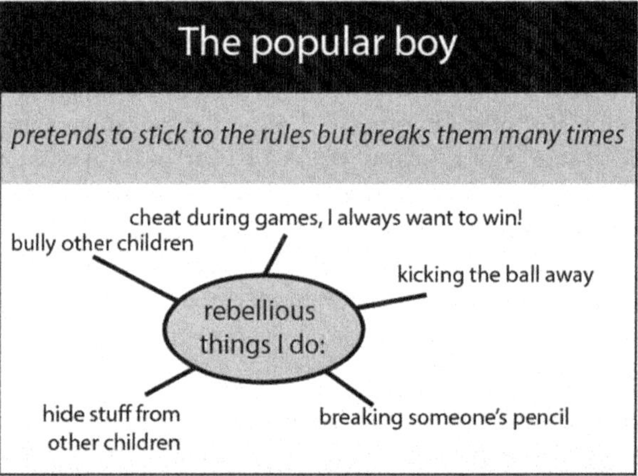
The popular boy
pretends to stick to the rules but breaks them many times
rebellious things I do:
cheat during games, I always want to win!
bully other children
kicking the ball away
hide stuff from other children
breaking someone's pencil

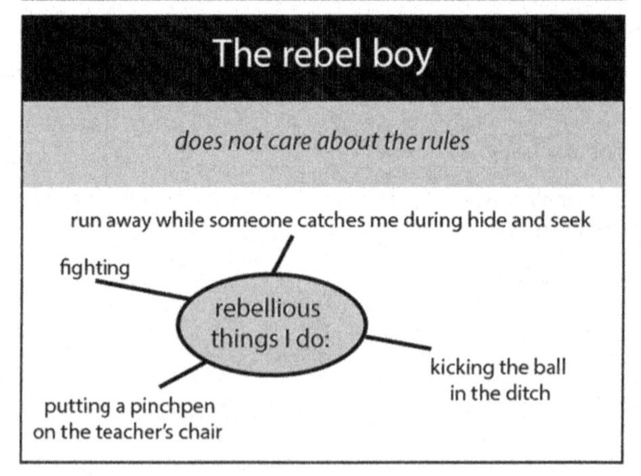
The rebel boy
does not care about the rules
rebellious things I do:
run away while someone catches me during hide and seek
fighting
kicking the ball in the ditch
putting a pinchpen on the teacher's chair

◀ **Fig. 15.2** Strategies of rebelliousness developed with a group of children 9–11 years of age in The Netherlands; by Britt Luijpers, M.Sc. student Delft University of Technology

> 'I don't like it that sometimes the other children don't want me to play with them.'
>
> 'When someone is really annoying, I push him in the bushes.'
>
> 'It would be great if someone would cheat on Bob. But I think he would laugh about it.'

Discussing ways of transgressing rules in a non-judgemental way provided insight into children's reasoning around their rebellious acts. One resulting game concept was 'Hide and Seek with a Twist.' Instead of one base point at which players can free themselves, there are four bases that sound when activated. Each player has a 'cheating device' and can use it only once to change the active base. There are different ways of cheating with this device. Some reveal the cheater instantly and others don't. A player can activate a base near to his or her own position in order to free oneself; prevent others from freeing themselves by de-activating the base they are trying to reach; activate a base to reveal someone's position to the seeker; move the base randomly to distract the seeker; or not use the cheating device. This design also 'allows' cheating like bluffing games, but it is up to the player to choose if and how he uses the cheating tool. The best strategy depends on the dynamics of the game but also on the preference of the player to be detected or not.

The Motivational Approach—Social Risk Taking

Rebelling means taking social risks. Individuals differ in their willingness to take those risks and in their experience of danger. For physical risk taking, this has been described in detail by Apter (2007a, b). In a playful motivational state, rules are broken to increase excitement and challenge the experience of being under control. In the context of social risks, the danger experienced is that of social repercussions. In a play context, teasing someone would be a typical example of social risk taking. Some children will be keen to engage in this while others will not. As for physical risks, the willingness of people to take social risks depends on their personal experience of control and protectedness. Apter (2007a) describes 'protective frames' which indicate varying levels of subjectively experienced (not necessarily objective) control and safety (see Fig. 15.3). In the detachment frame, someone only feels under control and safe, when not directly involved. An example in the realm of rebellion would be children's excitement about the adventures of rebel characters like Robin Hood or Pippi Longstocking.

The safety zone describes the experience of control and safety in situations which keep possible consequences of risk taking to a minimum. Toys can provide such a safety frame by, in a way, 'taking the blame' for non-conformist actions which are explicitly supported by the design. One of the oldest examples is the farting cushion. Placed on someone's chair it will make farting noises as soon as someone sits on it. It is particularly interesting for young children to bring adults into a situation in which they are seemingly breaking the conventional rules. Its being an 'official toy' and the possibility of sneaking it unseen onto someone's seat make it less likely that children have to endure negative consequences. The choice of person to play this joke on is in itself an interesting act of social judgement and provides the player with the opportunity to calibrate the amount of social risk and excitement. The proposed 'Hide and seek with a twist' provides a more modern version of designs that protect children from negative consequences of non-conformist actions. Within game play,

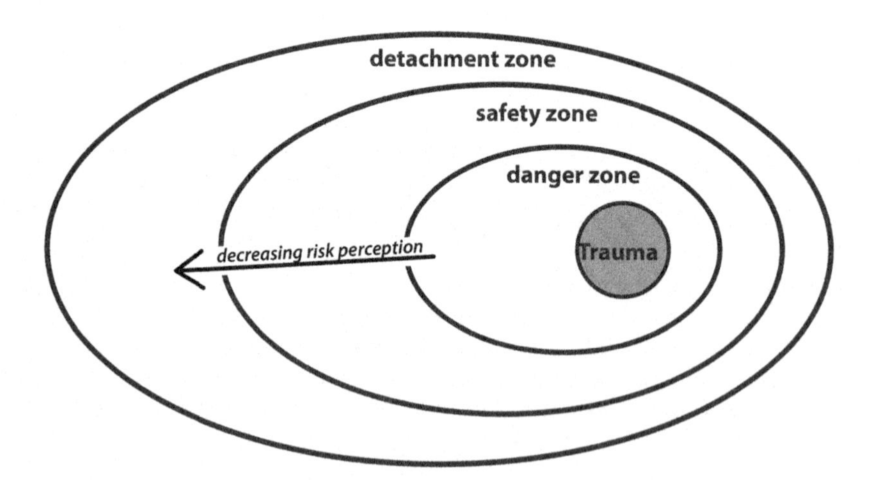

Fig. 15.3 Protective frames as described by Apter (2007a)

the use of a Joker to put other players in a worse position would be another example of social risk taking in a safety frame.

Acting in the confidence frame, means to embrace what is experienced as dangerous, trusting one's own capabilities and equipment to provide control and safety. An example of taking a social risk in the confidence frame would be to feel safe when play fighting with an unfamiliar child, or teasing someone to a degree that borders on bullying. Acting in the trauma zone means to feel safe in situations which put one's life at risk. In a social risk context, this could be play by engaging in deviant behavior like seriously harming others, which can lead to imprisonment. For a detailed discussion see Apter (2007b).

Protective frames are subjective experiences rather than objective facts. How safe one feels differs among people, situations, competencies, and values. Design ideas incorporating protective frames should therefore allow players to choose different levels and types of risk and control (see Van Leeuwen et al. 2012 for design examples). Applying it to the social risk domain would be particularly interesting in design for age 10 and above. Developmental trajectories from middle childhood to early and later adolescence show changes in priorities and values (LaFontana and Cillessen 2010) which have an effect not only on the frequency of non-compliant behavior, but also on who is rebelled against and what types of risks (social, physical, emotional) are taken. For example, young adolescents have a tendency to prioritize popularity in and conformity to a peer group, sometimes even if it compromises close friendships and health. This fact does not automatically lead to morally bad behavior, it characterizes a necessary developmental shift to gain respect and support from peers and secure belonging. In late adolescence this trend reverses and friendships are least likely to be jeopardized. Playful exploration and juxtaposition of types and levels of social risk could be a topic around which design for play could evolve for this age group.

Games presenting different degrees of social risk and different types and complexities of consequences can provide challenging and exciting play experiences. In addition, children could play in personally comfortable levels of risk and explore the reality of higher or lower risk levels.

Conclusion

Design for rebelliousness in the first instance sounds like a contradiction in terms—if something is designed for rebellion, then one conforms to the design by rebelling in a prescribed way. Real rebellion is to negate intended use, a practice children are particularly good at. So, should designers forget about it? This chapter intends to open a debate about the possibility and the need of design for rebelliousness.

Rebelliousness has been explored in the psychological frameworks of development and of motivation. Removing the exclusively 'bad behavior' stigma, it emerged that rebellion, seen as coping with social pressure (Apter 2007a), is a characteristic of everyday life which deserves attention. Competencies in decisions to conform or not, possible contexts for rebellion (the 'What' of rebellion) as well as strategies for expressing it (the 'How' of rebellion) have been shown to have relevance in the everyday life of children and therefore carry potential as starting points for design in the context of play.

While rebellious characters in children's stories are loved and embraced, the existing incorporation of rebelliousness in active play contexts largely mirrors the existing bias towards 'right' and 'wrong.' Breaking of rules is only allowed if the rules of the game or the overt design of a toy permit morally wrong behavior such as cheating. They seldom put the player in the position to make decisions about what rules to break or not. Neither are strategies of rebelling addressed.

Two basic design strategies, one implicit and the other explicit, emerge from exploring the 'What' and the 'How' of rebellion. The implicit one is directed at creating complex situations for free play, which grant autonomy to children to regulate their own affairs. Instances of contradictory intentions, the need for sharing resources, and the use of power, will in a self-organizing way create and regulate rebellious behavior greatly, due to the overriding motivation to keep playing.

An explicit strategy could directly invite rebellious play in various ways.

- Take the blame scenarios: meet children's needs to be non-conformist by providing 'outlets' to act in ways which are attractive but forbidden outside the provided play frame.
- Dilemmas: play/game scenarios, which confront players with dilemmas that force decisions to conform or not to conflicting requests. Dilemmas can be based on conflicting goals, power relationships,

and/or group affiliations. Protective frames should let players choose the amount of social risk they dare to take.

- Rule production: change and explore systems of rules—for example, choosing rules to break and replacing them with alternative ones.
- How to resist: explore what it takes to change the opponent's mind with respect to a request, and accept a conflict.
- Design strategies can be informed by developmental changes in social values and competencies. Motivational dynamics enable the imagination of different types of rebellion as well as to understand the needs for both conforming and rebelling. While psychological concepts can point to possible basic types of rebellious interactions in play, the everyday life of children, their needs and interests in rebelling have to be considered when deciding about concrete contents.

A designer's own attitude to non-conformism will inform the rules of such a game to a great extent. Independent of a chosen strategy, the design can allow one to succeed in the game only when never breaking rules (an exclusively moral stance) or permit breaking of only certain types of rules (e.g. differentiating between moral and conventional rules). Rebellious games can be predetermined or radically open-ended. These decisions mirror attitudes, interests and power relationships. Recognizing and supporting rebellion as a constructive force by design for play is a task for non-conformists.

References

Apter, M. J., & Smith, K. C. P. (1976). Negativism in adolescence. *The Counsellor, 23*(24), 25–30.

Apter, M. J. (1982). *The experience of motivation: The theory of psychological reversals*. London: Academic Press.

Apter, M. J. (2007a). *Reversal theory: The dynamics of motivation, emotion and personality*. Oxford: Oneworld.

Apter, M. J. (2007b). *Danger. Our quest for excitement*. Oxford: Oneworld Publications.

Bateson, G. (1972). A theory of play and fantasy. In *Steps to an ecology of mind* (pp. 177–193). New York: Ballantine.

Bergström, G. (2005). *Good night, Alfie Atkins*. Stockholm: Rabén & Sjögren.

Broadhead, P., & Burt, A. (2012). *Understanding young children's learning through play: Building playful pedagogies*. London: Routledge.

Caillois, R. (2001/1958). *Man, play and games.* Chicago: University of Illinois Press.

Cote, J. E., & Levine, C. G. (2015). *Identity formation, agency, and culture: A social psychological synthesis.* Oxford: Psychology Press.

Erikson, E. H. (1968). *Identity: Youth and crisis.* New York: Norton.

Fontana, D. (1988). Self-awareness and self-forgetting: Now I see me, now I don't. In M. J. Apter, J. H. Kerr, & M. P. Cowles (Eds.), *Progress in reversal theory* (pp. 349–360). Amsterdam: Elsevier.

Gielen, M. A., & van Leeuwen, L. (2013). Rebel by design: The merits of rebellious play and how to design for it. In *IASDR 2013: Proceedings of the 5th International Congress of International Association of Societies of Design Research, Tokyo, Japan, 26–30 August 2013.* International Association of Societies of Design Research.

Goffman, E. (1986). *Frame analysis: An essay on the organization of experience.* Boston: Northeastern University Press.

Gordon, G. (2008). What is play? In search of a universal definition. *Play and Culture Studies, 8,* 1–21.

Griffin, M., & McDermott, M. R. (1998). Exploring a tripartite relationship between rebelliousness, openness to experience and creativity. *Social Behavior and Personality: An International Journal, 26*(4), 347–356.

Grusec, J. E., & Kuczynski, L. E. (1997). *Parenting and children's internalization of values: A handbook of contemporary theory.* Wiley.

Helwig, C. C., & Turiel, E. (2011). Children's social and moral reasoning. In P. K. Smith & C. H. Hart (Eds.), *Wiley-Blackwell handbook of childhood social development* (pp. 567–583). London: Wiley.

Johnson, J. E., Welteroth, S. J., & Corl, S. M. (2001). Attitudes of parents and teachers about play aggression in young children. *Play and Culture Studies, 3,* 335–356.

Kuczynski, L., & Kochanska, G. (1990). Development of children's noncompliance strategies from toddlerhood to age 5. *Developmental Psychology, 26*(3), 398–408.

Kushner, D. (2007). Playing dirty. *IEEE Spectrum, 44*(12), 32–37.

LaFontana, K. M., & Cillessen, A. H. (2010). Developmental changes in the priority of perceived status in childhood and adolescence. *Social Development, 19*(1), 130–147.

Le Saux, A. (2011). *La boîte des papas* (The box of dads). Paris: L'Ecole des Loisirs.

Lee, J. J., & Hoadley, C. M. (2007). Leveraging identity to make learning fun: Possible selves and experiential learning in massively multiplayer online games (MMOGs). *Innovate: Journal of Online Education, 3*(6), 5.

Leeuwen, L. van, Gielen, M. A., & Westwood, D. (2012). Controlling experience or experiencing control? Reversal theory & design for play. In *Out of Control: Proceedings of the 8th International Conference on Design and Emotion, London, UK, September 11–14.*

Lefebvre, H. (1996). *Writings on cities* (vol. 63, no. 2). Oxford: Blackwell.

Luijpers, B. (2013). *Rebels at play - Exploring co-research to design a rebellious playground*. TU Delft: unpublished master thesis.

Maxwell, L. E., Mitchell, M. R., & Evans, G. W. (2008). Effects of play equipment and loose parts on preschool children's outdoor play behaviour: An observational study and design intervention. *Children Youth and Environments, 18*(2), 36–63.

Mireault, G., Rooney, S., Kouwenhoven, K., & Hannan, C. (2008). Oppositional behaviour and anxiety in boys and girls: A cross-sectional study in two community samples. *Child Psychiatry and Human Development, 39*(4), 519–527.

Nicholson, S. (1973). The theory of loose parts. *Man/Society/Technology—A Journal of Industrial Arts Education, 32*(4), 172–175.

Norman, N. (2003). *An architecture of play: A survey of London's adventure playgrounds*. London: Four Corner Books.

Piaget, J. (1997/1932). *The moral judgment of the child*. New York: Simon & Schuster.

Salen, K., & Zimmerman, E. (2004). *Rules of play: Game design fundamentals*. MIT Press.

Schaffer, H. R. (2006). *Key concepts in developmental psychology*. London: Sage.

Smetana, J. G. (2006). Social-cognitive domain theory: Consistencies and variations in children's moral and social judgments. In M. Killen & J. Smetana (Eds.), *Handbook of moral development* (pp. 119–153). Erlbaum.

Sonnentag, T. L., & Barnett, M. A. (2013). An exploration of moral rebelliousness with adolescents and young adults. *Ethics and Behavior, 23*(3), 214–236.

Spitz, R. A. (1957). *No and yes: On the genesis of human communication*. Madison, CT: International University Press.

Stevens, Q. (2007). *The Ludic City. Exploring the potential of public spaces.*

Sutton-Smith, B. (2009). *The ambiguity of play*. Cambridge, MA: Harvard University Press.

Sutton-Smith, B., Mechling, J., Johnson, T. W., & McMahon, F. (2012). *Children's folklore: A Sourcebook*. Routledge.

Toscano, A. (2008). In praise of negativism. In S. O'Sullivan & S. Zepke (Eds.), *Deleuze, Guattari and the production of the new* (pp. 56–67). London: Continuum.

Turiel, E. (1983). *The development of social knowledge: Morality and convention*. Cambridge: Cambridge University Press.

Turiel, E., & Wainryb, C. (2000). Social life in cultures: Judgments, conflict, and subversion. *Child Development, 71*(1), 250–256.

Turiel, E. (2006). The development of morality. In W. Damon & R. M. Lerner (Series Eds.) & N. Eisenberg (Ed.), *Handbook of child psychology: Vol. 3. Social, emotional and personality development* (6th ed., pp. 789–857). Hoboken: Wiley.

Turiel, E. (2003). Resistance and subversion in everyday life. *Journal of Moral Education, 32*(2), 115–130.

Turkle, S. (2005). *The second self: Computers and the human spirit.* Cambridge, MA: MIT Press.

Van Leeuwen, L., Westwood, D., & Gielen, M. (2012). Controlling experience or experiencing control? In J. Brassett, P. Hekkert, M. Ludden, J. McDonnell & M. Malpass (Eds.), *Proceedings of 8th international design and emotion conference London 2012.* ISBN 978-0-9570719-2-6.

Waterman, A. S. (1982). Identity development from adolescence to adulthood: An extension of theory and a review of research. *Developmental Psychology, 18*(3), 341.

Williams, D., Kennedy, T. L., & Moore, R. J. (2011). Behind the avatar: The patterns, practices, and functions of role playing in MMOs. *Games and Culture, 6*(2), 171–200.

Author Biographies

Lieselotte van Leeuwen is a senior lecturer at HDK-School of Design and Crafts, University of Gothenburg, Sweden.

Mathieu Gielen is an Assistant Professor in Design for Children's Play at TU Delft, Faculty of Industrial Design Engineering, The Netherlands.

CHAPTER 16

Hong Kong PolyPlay: An Innovation Lab for Design, Play, and Education

Rémi Leclerc

Introduction: Background and Origins

Toy Play

In 2004, the Hong Kong Polytechnic University School of Design (SD) opened a cluster of design research laboratories, among them the Toy Design Lab. The latter recruited faculty members with years of toy design industry experience, and aimed at investigating the realities of the toy manufacturing industry, one of Hong Kong's (HK) top three export industries in terms of revenue for the Special Administrative Region. The Lab was intended as an inspirational platform to develop creative design for play strategies, and dedicated itself to fostering from within HK a culture of innovation that would embrace creativity, technology, and communication, so as to support HK's toy, entertainment, and educational product industry in maintaining its global predominance as a leading innovation, development, and marketing hub for play products.

R. Leclerc (✉)
The Hong Kong Polytechnic University, Hung Hom, Hong Kong
e-mail: remi.leclerc@polyu.edu.hk

L. Magalhães and J. Goldstein (eds.), *Toys and Communication*,
https://doi.org/10.1057/978-1-137-59136-4_16

From Toy to PolyPlay

In 2007, the unit was renamed PolyPlay, to reflect the fact that projects carried out under its aegis had extended its initial research focus beyond the confines of toy design: a natural shift from design for play to … play for design, or Design Play, whereby reflective thinking on processes acquired in the design for play practice led to seeing the relevance of play theory and toy design practices to other fields of design. PolyPlay acts as a research lab, an education resource facility, a professional consultancy, and a design development office. Its investigation work focuses on the overlap of play, design, and education, and it transfers knowledge thus accrued to broader innovation contexts. PolyPlay's consultancy, research, licensing, and entrepreneurship projects feed SD educational experiences. Expertise gained in experimental projects is shared with academic, community, and industry stakeholders.

PolyPlay Research

Research at PolyPlay is anchored in the exploration of different notions of play, and the conventions of interactivity, and draws on characteristics of both traditional and contemporary play and other objects to develop prospective play, educational, recreational, and interactive products, environments and multimedia systems. Strong from experiences gathered in these areas, PolyPlay works at explicating the links existing between play, design, and education, to enrich such fundamental topics of study as design thinking, strategic design, design semantics, user-centered design research processes, and design communication.

PolyPlay Web

A compendium of works generated by the facility since 2007, polyplay.hk (Leclerc 2014) showcases a kaleidoscope of design for play projects, research chapters, conference and seminar presentations, consultancy work, Work-Integrated Education (WIE) projects, and exhibitions. Curating a natural emergence in various trends of play, the digital archive showcases over 150 papers, presentations, exhibitions, and 'playworks,' the latter categorized along 9 branches of play, following an organic, 'bottom-up' classification of design outcomes generated at SD over the years. Selected playworks highlight the different approaches to toy design applied in student projects, which promote user-(as-player)-centered experimental and critical forms of design processes. Homo Sapiens sic Homo Faber sic Homo Polyludens.

PolyPlay Conceptual Framework

Play, Culture, and Design: Conceptual Bridges

Play is seen by developmental scientists as being intrinsic to humans' biological (individual, psycho-physical) and social (group, cultural) development, through physiological, social, and cognitive stages, as preparation for life, a condition for welfare and survival, or even evolution. Yet many recognize the challenge in substantiating its functional significance. For Huizinga (1938) play is an agent of civilizational or cultural development. Building on Huizinga, Caillois (1958) offered a variable geometric model of play, articulating four fundamental play forms (agon [competition], alea [chance], mimicry [imagination], and illinx [vertigo/sensory distortion]), to be placed individually or combined, on a 'ludus-paidia' continuum opposing rule-governed, skill-requiring game play, to unstructured free play, so as to define ludic activities and explain human behavior. He saw the continuum as metaphor characterizing the instability of cultural dialectics, as in time paidia ultimately yields to ludus, with control giving way to freedom and again yielding to control, in a historical pendular conversation transforming culture. There lie useful parallels to be drawn between divergent (exploratory, creative, expanding, 'right brain'—ludus) and convergent (discriminating, analytical, integrative, 'left brain'—paidia) iterative dialectics observed in design processes.

Lamenting scholars' usage of play as 'a persuasive discourse or implicit narrative [...] by members of a particular group affiliation or discipline to lend validity to their beliefs and interpretations,' Sutton-Smith (1997) enunciates seven rhetorics of play so as to reveal play's 'ambiguity'—while admitting he had yet invented another form of rhetoric of progress: *biologically*, play activities function to 'reinforce the organism's variability in the face of rigidifications of successful adaptation. [...] This variability covers the full range of behavior from the actual to the possible'; and '*psychologically*, as a virtual simulation characterized by staged contingencies of variation, with opportunities for control engendered by either mastery or further chaos. Clearly the primary motive of players is the stylized performance of existential themes that mimic or mock the uncertainties and risks of survival and, in so doing, engage the propensities of mind, body, and cells in exciting forms of arousal.' He concludes his review by summarizing play as 'potentiation of adaptive variability.' This interpretation of play offers an exciting perspective to enrich appreciation of design's relevance to society, as an agent of cultural transformation.

Toys and Communication? Toying with Design!

The creative act of design is a progressive form of discourse—*a line making a point*, a form of purposeful play: literally, design *shapes* culture. If, as Huizinga contends, culture is the outcome of play, then play and design share similar functions of cultural agency.

In *Emile*, Rousseau (1762) suggested play facilitates the acquisition of competencies (thinking, that is, potential) and applied skills (making, that is, know-how), both core to design. Inspired by Rousseau, von Schiller (1793) claimed play was the only opportunity for humans to fully develop their humanity by setting free the two aspects of its double nature: sensation and thought ... in other words, the left-brain/right-brain complementarity characterizing design thinking. The formative value of Kindergarten's creator Froebel's proto-constructivist 'Gifts and Occupations', created in 1837, (Froebel 2005, 2015), basic toys to nurture children's learning through play, were acknowledged by artists, designers, and architects such as Kandinsky, Gropius, Klee, Lloyd Wright, and Fuller. Montessori, whose Method (1912) is now applied in schools all over the world, conceived the 'Casa dei Bambini' (Children's House), in which afternoon design activities were scheduled as part of an innovative curriculum. Rand (1965) made a case for play in design education, noting that 'a problem with defined limits, implied or stated disciplines which are, in turn, conducive to the instinct of play, will most likely yield an interested student and, very often, a meaningful and novel solution.' In 1977 Munari established his 'Giocare con l'Arte' (Playing with Art) educational method based on Montessori's, advocating 'non dire cosa fare ma come' ('not to say what to do, but how') and *observing*, *making*, and *reflecting on the making*. Manu (1995) suggested that 'beyond its environmental imperative sustainable design will have to be engaging, challenging, rewarding, absorbing, non-frustrating, and of repeat experience value'—as in play. For IDEO president Brown (2008) designers should take play seriously. Kristiansen and Rasmussen (2014) utilize Lego elements and narrative play to get corporate executives to create props for creative strategic business development sessions. Whack

Pack creative cards author von Oech (1983) puts it best: reasonably thumbing his nose to Plato, he argues that 'necessity may be the mother of invention, but play is certainly the father.'

How Does the Design Process Play Out? Visualizing Design Play

How are design processes similar to play and why does it matter? Contemplating their uncertain and immanent nature, from their fuzzy front end and 'wicked' evolution, to the harmonization of creative tensions and generation of an appropriate solution, one may see similarities with the spontaneous, ambivalent, 'magic circle' of ludic experiences. Riding play as a 'friendly Trojan hobby horse' to enhance design education and explicate design's relevance to culture, these similarities may be brought to light by juxtaposing essential play and design concepts: process and tools. Design projects are characterized by a two-dimensional temporality, with (1) the 'diachronic' unraveling of a process, akin to narrative play, and (2) projects following a planned, yet evolving course of actions, deploying successive 'synchronic' activities, methods, or tools, which are similar to play activities. The former is represented as a continuum (Fig. 16.1), the latter in a taxonomy (Table 16.1).

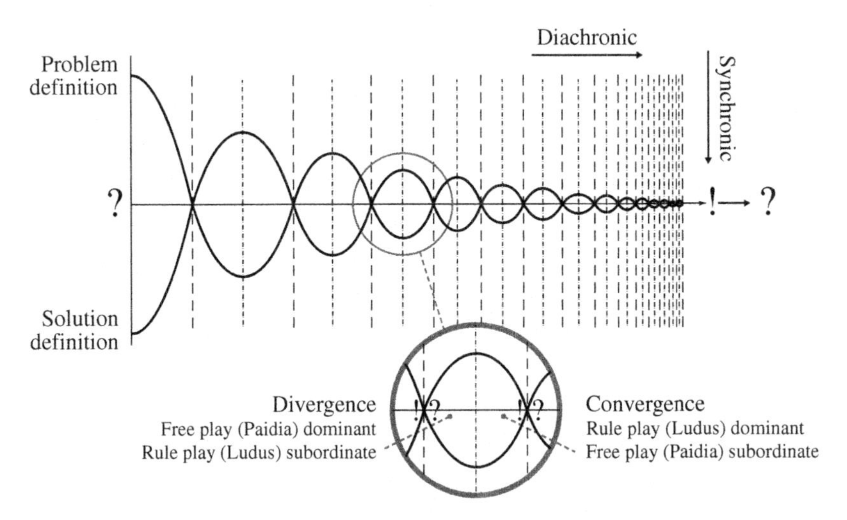

Fig. 16.1 Design Play Continuum, Leclerc and Wan (2012)

Table 16.1 Design play taxonomy

	Forms of activity	*Play activities*	*Design tasks*
Play & design	**Competition** Pushing/challenging oneself/others, sensory distortion, testing limits, subversion, empowerment	Contests Games Games of strategy Completing a game Achievement	Competitions Project bids Design strategy Completing a project Promotion
	Cognition Learning, defining problem-solving, rule-based game, strategy	Games of strategy Quizzes & puzzles Science kits Tinkering experiments Outdoor exploration	Strategic design Problem-solving Reverse-engineering Research through design Field studies
	Creative Exploration, expression, arts, aesthetics	Drawing Tomfooling Arts & crafts Performance art	Visualizing Brainstorming Drawing & model making Presentation & showmanship
	Imagination Self-social, fantasy, pretending, mimicry, narratives	Story telling Role play Make believe Costume play	Scenario building User empathy Prospection & simulation Fashion & identity building
	Manipulation Fine motor, constructing, making	Model making Construction toys Arts & crafts Tools & stationery	Model making, prototyping Design principles Craftsmanship Tools & machine tools

(continued)

Table 16.1 (continued)

Forms of activity	*Play activities*	*Design tasks*
Sensory Motor Perception & body, feeling and moving, thrill, pleasure	Sports Sensory play Emotional experiences Altering perception Pleasure	Ergonomics, body storming Gestalt, Kansei, synaesthetic Experience design Blindfolding Aesthetics
Chance Gamble, weighing the odds	Chance games Gambling Magic	Gambits Opportunity & venture Delight

The Design Play Continuum (Leclerc and Wan 2012): simultaneously defining problem and solution, designers alternate between (1) divergent phases of discovery, where creative 'fun' play—Paidia—dominates; and (2) convergent, analytical phases of synthesis, during which rule-based skill-requiring decision-making 'serious' game play—Ludus—prevails. Design projects thus iteratively play out on a dialectical co-evolving problem-solution continuum, refining a proposition until the formulation of a socially and culturally 'appropriate' solution is harmonized in response to a brief.

The Design Play Taxonomy (Leclerc and Wan 2012) mirrors examples of play and design activities to highlight design's inherent play characteristics. The structure follows a 'bottom-up' developmental pattern, from chance at its base (characterizing our conception), to competition at its top (strategic integration of skills), through sensory motor, manipulation, imagination, creative, and cognitive play. Synchronic design activity examples cited in the Taxonomy are actions, or tasks chronologically performed in sequence, in parallel, or repeatedly, in design's diachronic process, represented by the Design Play Continuum.

Play's Rightness, Design's Appropriateness

Huizinga and Cross: Designerly Ways of Playing, Playful Ways of Designing

It is often said that to play is to live, or that life is a play ... that said, nothing's said. Defining play is as challenging as defining design; the former characterized by ambiguity, the latter by uncertainty. While the conversation between sensation and thought, or feeling and reasoning, is core to play, so it is to the design practice: each activity involves learning by doing, an experiential way of knowing. It is core to design in all its forms, be it in Practice, Exploration, or Study—the triangular scope of design activities described by Daniel Fallman (2008)—to which must be added Education in order to square it right.

If play is a generator of culture, is it a useful instrument for improving design practice, exploration, study, and education? As design shapes culture, its processes are characterized by uncertainty and immanence. Parallels could yet again be drawn with play as agency for culture. For Sutton-Smith (1997) play's quirkiness (challenging conventions), redundancy (trying out scenarios), and flexibility (adapting to circumstances) might nullify the rigidity (a tendency to be satisfied with *the* solution) that sets in after successful adaptation (or designs), thus enhancing human's adaptive variability.

Huizinga describes play as happening in a 'magic circle,' a place and a time outside ordinary boundaries. In design this place in the project, where designers can relate to five essential play characteristics identified by Huizinga, as listed below:

First, play is a relatively free, voluntary activity. Caillois (1958) further suggests that any play activity that would be forced onto an individual would cease to be play. Intrinsic motivation, or proactivity, is essential to design, and so is the lifting of boundaries to access possibilities: in creative design, innovation is limited only by one's imagination.

Second, 'play is distinct from "ordinary" life; rather, it is stepping out of "real" life into a temporary sphere of activity with a disposition all of its own'; and so design projects engage in speculative scenarios, abiding by exotic rules regulating imaginary representations of and relationships between body, place, time, objects, and images.

Third, play has its own time and space, limited and secluded, 'it is "played out" within certain limits of time and place. It contains its own order course and meaning'; as in the design project, a 'magic circle' suspended in time where stakeholders are given the freedom to engage with existing situations and define preferable ones, enacted through creative, narrative, and inaginary design processes.

Fourth, play creates order, is order, and demands order absolute and supreme. It abides by certain degree of rules yet promotes transgression and disorder. Huizinga writes:

> The profound affinity between play and order is perhaps the reason why play, as we noted in passing, seems to lie to such a large extent in the field of aesthetics. Play has a tendency to be beautiful. It may be that this aesthetic factor is identical with the impulse to create orderly form, which animates play in all its aspects. The words we use to denote the elements of play belong for the most part to aesthetics, terms with which we try to describe the effects of beauty: tension, poise, balance, contrast, variation, solution, resolution, etc. Play casts a spell over us; it is 'enchanting', 'captivating'. It is invested with the noblest qualities we are capable of perceiving in things: rhythm and harmony.

Design, like play, abide by certain rules, yet promote transgression and disorder. In design this results in paradigm shifts or as Szent-Gyorgyi (1963) put it, discovery consisting of 'seeing what everybody has seen and thinking what nobody has thought.' In design, where what-could-be becomes what-ought-to-be, 'what if?' questions lead to prospective scenarios. Knowledge of design possibilities thus created, although feasible only in such contexts, informs iterative development and shapes desirable futures. Rittel (1964, in Rith and Dubberly 2007) explained that 'any theory of innovation including a theory of design must be based on a theory of action, not a theory of knowledge (epistemology) alone. [...] Design is concerned with instrumental knowledge (how what-is relates to what-ought-to-be), how actions can meet goals. The process of argumentation is the key and perhaps the only method of taming wicked problems.'

Fifth, play is distinguished from routine affairs by its absence of material consequences. 'Play is connected with no material interest, and no profit can be gained from it.' This does not contradict the real-world characteristic of design: design fails not always for providing the wrong answer.

In the opening salvo of a series of three essays that led to the definition of 'designerly ways of knowing,' Archer (1979) suggested it was 'the collected experience of the material culture, and the collected body of experience, skill and understanding embodied in the arts of planning, inventing, making and doing.' This way of knowing is distinct from that which occurs in science and the humanities. It is placed on the top of a pyramid, suggesting design's primacy in its relation to the other two forms of knowing. For Cross (1982), designerly ways of knowing are concerned with 'appropriateness,' to be implemented as a 'third area' of education, next to science and humanities education, respectively concerned with 'truth,' and 'justice.' Cross identified five aspects of designerly ways of knowing:

- Designers tackle 'ill-defined' problems (players spontaneously embrace ambivalent, self-generating processes).
- Their mode of problem-solving is 'solution-focused' (whereas play's purpose lies in itself, players create order, harmony, in a transformative act).
- Their mode of thinking is 'constructive' (as we play, we do; as we do, we learn).
- They use 'codes' that translate abstract requirements into concrete objects (players engage in symbolic representation and communication).
- They use these codes to both 'read' and 'write' in 'object languages' (as they draw on symbolic meaning, players transform time and place).

From these ways of knowing, he drew three main areas of justification for design in general education, which echo play-for-education proponents' tenets: design develops innate abilities in solving real-world, ill-defined problems. It sustains cognitive development in the concrete/iconic modes of cognition. It offers opportunities for development of a wide range of abilities in nonverbal thought and communication.

This epistemological view of design is promoted as an integrative way of acquiring knowledge and generating meaning while remaining attuned to the immanence of contemporary society. Taking the point further, one could argue that it is *informed by* and *informs all* human activity. For Simon (in Heskett 2008), design is not restricted to making material artifacts, but is a fundamental professional competence extending to policy-making and practices of many kinds and on many levels: again for Simon (1969), 'everyone designs who devises courses of action aimed at changing existing situations into preferred ones.' Gero (1990) suggests that 'designers are change agents', engaged in what Lasn (2006) refers to as the 'generation of meaning': *design is a line making a point.*

Huizinga's description of play bears similarities to design project processes: play presupposes culture (design shapes culture), does this as a 'significant form' (as a 'line making a point') with a 'social function' (design serves society), is based on the manipulation of certain images and a certain 'imagination' of reality (that is, its conversion into images—design uses visual codes), and an activity is play if it includes elements of uncertainty (design 'embraces' uncertainty; for solutions have by definition yet to be imagined). In play the doing and the thinking occur seamlessly; in design defining a problem and defining a solution occur simultaneously. Practice is the amalgam of the two.

To design play brings the allure of freedom, the excitement of creativity, the romantic appeal of imagination, the enticement of discovery, the humanism of communitas and social inclusiveness, the balance between individualism and socialization, the healthy facetiousness of subversion needed to innovate, and the purposeful depth in generating meaning and bring order to chaos. 'The playing adult steps sideward into another reality; the playing child advances forward to new stages of mastery ... Child's play is the infantile form of the human ability to deal with experience by creating model situations and to master reality by experiment and planning' (Erikson 1950).

Design for Play Is a Prototype for Design

'The toy is children's initiator to art,' mused Baudelaire (1853). While the apparent uselessness of toys seems obvious to many (toys are not meant to perform any task), they are symbolic tokens reflecting understanding of society, and act as culture mediators. Just as artworks, far from being useless, toys are important to humans—and could be seen as:

1. Miniature replicas of objects given to children to introduce them to the ways of the world;
2. Tools aiding the development of sensory, physical, social, emotional, and cognitive abilities;
3. Props for physical, imaginary, fantasy, creative, cognitive, and/or aimless play;

4. Ritualistic, or symbolic objects owned or shared, collected for use or display in private or public environments, and providing psychological, social and emotional order;
5. Any object made or found, on which a human will project meaning for purposeful or aimless play, the interaction with which results in a transformation of the self.

While toys, or playthings, are designed objects, their rich, symbolic, yet ambivalent nature, also places them in the realm of the artistic experience. For historian Léo Clarétie 'the toy industry embraces in whimsicality and fantasy the best of all crafts combined.' Knowledge of form, color, texture, materials, and manufacturing processes, and from fields as diverse as psychology, the arts, language, education, communication, and marketing, shape toy design practice and enrich design thinking curricula for students of all design disciplines. Hence the relevance of toy design practices to broader design contexts.

To paraphrase Saul Bass, 'toy design is play made tangible.' Sutton-Smith (1986) made a case for toys as vehicles of culture in modern American society and the paradoxes regarding interpretations of various social contexts. A historical analysis reveals forces underlying values pertaining to four toy contexts and eight ideological interpretations:

- *Family*, where the toy emerges as a mediator of such concepts as bond and obligation, solitariness, and consolation.
- *Technology*, with the toy-as-machine, focusing on the intrusion of Information Technology, through media such as video games, technology of toys for infants, and toys as tools for infants.
- *Toy as education*, exploring the toy as achievement and the merging of play and work in children's early years.
- *Toy-as-market* reveals novelty and agency, how toys are idealized.

He also describes three kinds of toys: toys of acquaintance, toys of age- and sex-stereotypes, and toys of identity.

The relevance to design thinking is multiple as it uses the familiarity of the toy—an object close to young design students—to discuss aspects of semantics and identity within consumer culture.

Toy Design Is a Comprehensive, 'Radical' Form of Design

In his 2003 book *The Blockbuster Toy! How to Invent the Next Big Thing* Del Vecchio charts a 'blockbuster' Toy Ideation Matrix, a framework articulating essential elements for toy design which could be applied to broader product design practices. The matrix lists factors to consider in the design of a successful toy. Designers are invited to mix and match options from each of the Matrix's factor categories. It is shared in design thinking as a tool for students to develop their projects. Its usefulness to design is thus:

- *Toy Categories* (e.g. Infant and Preschool, Dolls, Action Figures, Video Games, etc.): an appreciation of product classification useful for identifying project opportunities.
- *Child Emotional Needs* (e.g. Pride, Accomplishment, Sensory Gratification, etc.) and *Parent Emotional Needs* (e.g. Child's Happiness, Child's Safety, Child's Health, etc.): an appreciation of multiple stakeholder needs enriching empathic user-centered design research—the ability to feel as users do.
- *Transformation* (i.e. toy play leading to development profiles such as Master, Creator, Emulator, etc.): awareness of play value's effect, here the development of psychological profiles nurtured through object interaction. Such awareness facilitates rationalization of product value.
- *Trends* (e.g. KGOY—'Kids are Getting Older Younger', World Awareness, 'Just for Me' etc.,): to emphasize importance of inscribing design in a broader socio-cultural context.
- *Marketing* (e.g. Public Relations, Advertising, Sales Productions, etc.): to raise awareness of the pragmatics of effective market communication.

The appreciation of toy design processes and factors enriches design thinking: while play theory's relevance to design is made apparent through the Design Play framework, the richness of toy design offers design at large a 'prototype' for practice; as it draws on design's best practices: user research, product interaction, semantics, narratives, cultural awareness, technological constraints, product communication, product safety issues, etc., thus vindicating Clarétie.

Ways to PolyPlay

PolyPlay is neither a science nor a whim. Call it a whimsical science. PolyPlay projects are informed by:

1. Contextual and theme mapping, to identify and define design opportunities, be these driven by personal motivations or latent market opportunities—this is game plan generation.
2. Product, market and technology research, to benchmark innovation opportunities, and communicate rationales through visual and text codes leveraging the power of information graphic narratives.
3. User-as-player-centered research probes, to extract a wish list of values and expectations from project stakeholders in the research phase of projects, so as to inform design concepts and tangible models for validation by the same stakeholders in subsequent development phases—because play intuitively fosters quality affordances.
4. Creative techniques borrowed from conceptual art and outsider art, and references made to the pioneers of pop art—to give user-centered approaches an inclusive human face.
5. Experimental ideation processes such as Hackshops (Leclerc 2010), to complement the 'discipline' of design with playful creative experiments tapping into the unconscious—because play is respectable only when it has a mind of its own, when it is left to its own devices, a voluntary from of creative participation and ontological development—where in creative design (especially creative toy design) discipline and rigor is enriched by aimless, free-form play.
6. Critical analysis of conventions of interactivity—how would anything be called play if not for that?
7. An appreciation of characteristics of both traditional and contemporary play, to re-contextualize timeless cultural patterns into current-day society—so that play remains new as time goes by, and that novelty never does wrinkle and never grows old.
8. Communication tactics borrowed from poetry, humor, and satire—to harness the emotional whimsy of the human experience.
9. Multimedia communication platforms—to engage designers in play experiences, which they will convert into rich immersive user narratives.

9.1/2 Nurturing personal design processes ...
9.3/4 ... to play with the industry's conventions and ensure socially appropriate innovation.

Playworks ...

... and there is no better way of feeling play at work than when one is working—or designing—for play. As one works for play, one never does: indeed one plays for work, demonstrating Sutton-Smith's point that 'the opposite of play is not work. It is depression.' Since its inception in 2004 as the Toy Design Lab, SD faculty members collaborating with what is now known as the PolyPlay lab have supervised hundreds of design for play projects, contributing to the creation of a repository of experiments the creative spark for which was mostly ignited by students. Over the years, students 'playworking' on projects supervised by lab collaborators have developed an appreciation of:

- Play history, theory, definitions, and categorizations
- Entertainment, education, recreation
- Child development: age grading, physiological, sensory, psychological, social, and emotional abilities
- Toy market research
- User research: context mapping and cultural probes
- Critical analysis of cultural conventions
- Technology and invention
- Interaction
- Play pattern, depth, and value
- Iterative toy design development and manufacturing technologies
- Toy safety
- Techniques for toy presentations and communication.

In most cases, students were given carte blanche and had to set their own project brief, a challenge often daunting for young designers, yet usually generating outcomes rich in results. Curating this natural emergence in various trends of play, the PolyPlay website showcases over 150 playworks, along 9 branches of play. The resulting accumulation of playworks eventually clustered around these categories represents an organic classification, of local youth's preferences and (mis-)representations of what

Desire / Pleasure
Taste Play
Music Play
Leisure Play

Wonder/Whimsy
Eco Play
Art & Science Play
Magic-A-Dime Play

Identity/Social
Culture Play
Character Play
Game & Media Play

Fig. 16.2 Playworks categories, created by author

play is about in an East Asian contemporary society. These may be grouped along three core concepts (see Fig. 16.2):

Identity and Social (Culture, Character, and Game and Media Play): here the projects stem from a desire to explore notions of self, and design narratives of social engagement.

Wonder and Whimsy (Eco, Magic-a-Dime, and Art and Science Play): the drive for discovery and enchantment, the fundamentally curious nature of design.

Desire and Pleasure (Taste, Music, and Leisure Play): the promise of design aesthetics' sensory rewards, and the emotional contentment lying in drifting and idleness.

There is no 'educational play' category: all toys facilitate some form of discovery, transformation, and development, including those used in aimless play. Experiential, constructivist, or didactic imperatives run throughout the categories.

Art and Science Play

For Baudelaire, toys initiate children to art. We could also see that toys initiate them to science. And magic—and then some. The best way to design an art or science kit is to play with art or science principles. The exploratory, creative, and developmental steps involved in designing an art or a science kit naturally foreshadows those which players will follow. Also as one manipulates instruments, tools, ingredients, and components, one gets to appreciate Albert Einstein's assertion that play should be seen as the highest form of research. Works featured in this section naturally merge both disciplines' characteristics through a whimsical game of mirrors, using art to stage science's wonderful nature, and science to substantiate art's fundamental aesthetic qualities. From Tom Tit (1843) to Arvind Gupta, art and science kits demonstrate the whimsy of natural phenomena—the tangible dream stuff all toys are made of. Note all three Tom Tit's 1890s' publications by Larousse were awarded the Medal of Honour by the Society for the Encouragement of the Good (Société d'Encouragement au Bien), while Arvind Gupta is internationally commended for his social innovation work.

Character Play

Play patterns generated by the reification of a popular myth are derived from an original narrative. Character toys are interfaces to a world of motivations, and mediate value; as such they combine toy with storytelling, inviting players to identify with role models, adapt adventures to their social needs, and act out their emotional development. In a global economy dominated by mediated images and fabricated narratives, characters are everywhere—no wonder then that children instantly connect with the many characters presented to them, and many young toy designers want to create character toys. PolyU Design students are no exception, and more than a third of the Playworks featured in this website are characters to play with. For toys to feature in this category, a figure needs to be staged in a specific story, purposely written for the toy, however short: with character toys, all it takes is visual personality and a narrative spark for children to engage in a world of fantasy. Projects often make reference to popular culture, movies, animé, experimental comic and children's books from French publishers *L'Association* or *Le Rouergue*, and Campbell's Hero's Journey as summarized in an internal Disney memo circulated by Christopher Vogler in 1985.

Culture Play

Design Schools are privileged spaces—where and when else would designers make the best of their youth, and play with material culture and the generation of meaning in a consumer society? From within the necessarily safe confines of a university, design students should be given the opportunity to revisit conventions of interactivity, affordances, identity, and everyday tasks, to formulate objects that 'talk' as much about us as they do about, and between, themselves. Toying with playful interactions, the projects featured in this category get as close to artistic interpretations of functionality as it is desirable in the context of functional design. Turning the tables on so-called 'pure design,' function here follows irony, poetry, humour, fantasy, or nonsense, while offering practical solutions to everyday needs. Reference is often made to the work on 'Critical,' 'Speculative,' or 'Ludic' Design by Dunne, Raby, and Gaver, or even Carelman's perspectives (1969) on 'unuselessness,' where design—not to be confused with art—is leveraged so as to engage society in considering our interactions with objects and their roles in our culturally transient postmodern contexts.

Eco Play

A great way to look at the toy industry is through the lens of its raw materials. Toy materials take millions of years to make, while toys themselves are at best played with for a few hours. More, it takes seconds to discard a toy, and a few thousand years for its materials to biodegrade and be absorbed back into the Earth's natural cycle. So much resource for products which are acquired through discretionary spending, and are qualified as superfluous, trite, or inessential … but then toys, like art, are essential, precisely because they do not serve tangible, easily defined purposes. This section showcases one of PolyPlay's preferred areas of innovation: the design of tools for play. Tools that process easy-to-find recyclable or biodegradable materials, such as molds to form sand or ice elements for construction toys, or cutting tools to process PET bottles. Concepts featured also include hardware which 'parasitically' upcycle found objects, such as clips and fasteners assembling everyday objects into playthings. A common feature running throughout these tools is that they facilitate the creation of toys that make-do with their environment, and 'disappear' after usage. The work of environmental educator Louise Chawla and the eco wheel strategy often frames and inspires design processes.

Game and Media Play

Are games play? If games involving dynamics of power are considered a corruption of play (winning or losing leading to players' dominance over other players), then how should they relate to play, given that the moment play involves power it ceases to be play? By definition games are platforms mediating social interaction: a game is a social medium. Today, much of games' social interaction is electronically mediated through online social networks. While games played alone are shared online through digital media, multiplayer games do not necessitate the physical presence of partners to engage in play ... other players become figments of one's imagination, or constructs carefully considered 'profiles.' Grappling with the complexity of contemporary social play, students designing for game play have slid design for play concepts on Caillois' (1958) Paidia/Ludus scale (a conceptual continuum opposing unstructured, spontaneous activities, or free play on one end, to rule-based structured game play on the other). Thus, they explore the possibility of structured yet progressive social interactive play, so as to foster appropriate symbolic and value exchange that nurtures benevolent and inclusive socio-cultural development. Fluxus' art, Salen and Zimmermann Rules of Play (2003), and Pearce's fundamental elements of game (2006) are used to structure game patterns.

Leisure Play

Have we forgotten how to play? The rise of a Leisure Society, much trumpeted in the 1970s, never came to be. The current global socioeconomic paradigm has people leading commoditized lifestyles, characterized by overwork and stress—especially in HK and China's major cities. Also, while the media and the fashion industries have reshaped sports into a spectacle, leisure activities have become branded affairs, thus defeating the purpose of leisure. In study trips and design for leisure projects, students were required to critically discriminate between play and work conventions, and contextualize the social, cultural, emotional as well as physical importance of recreation and exercise in modern society. They have ascertained the relevance of play to contemporary culture. Exploring emerging recreation and outdoor activities in China and the region, they created innovative solutions for the way people play and replenish, to lead healthy lives, create 'situations,' or simply escape from branded consumer 'experiences.' While students revisit conventions of sports games, they also turn to nineteenth-century urban romantic drifters, Poe's atmospheric depictions of twilight environments such as that in the 'Man of the Crowd' (1845), situationist psychogeography and *dérive* (drifting) tactics, or Punks' celebration of urban ordinariness such as that of The Jam's in *That's Entertainment* (1981).

Magic-A-Dime Play

At the root of all toy inventions lies a playful interpretation of natural phenomena: many toys mediate scientific knowledge under the guise of narrative and poetic license. Consequently, much wonder emerges from playing with the most elementary playthings. Gupta's online repository of easy-to-make Toys from Trash celebrates the magic one can find in the simplest of physical or chemical principles. Likewise, with as few materials as possible, designers can conjure pocket-sized magical trinkets, from one-time marvels to skill-building training tools, thus elevating the lesser tchotchke, gadget, premium, curio, or novelty, into whimsical ephemeral play wonders. One, two, pick up sticks. Three, four, toys galore. Learning to acquire the ability to conjure up magic for 10 cents should be part of every toy designer's training—or every designer for that matter—as simplicity is often the hardest thing to capture. This section gathers works students have produced illustrating this toughest of design challenges: simple play, maximum bang for the dime.

Music Play

While we compose visuals to admire, cook food to taste, concoct fragrances to smell, apply textures to touch, we *play* music. Music is as much a form of expression as it is an impression; it is an intentional arrangement of sounds we produce through the use of our body or instruments, for sharing emotions or ideas, resulting in an immediate appreciation of these. Music is soothing, entertaining, pleasurable, as well as violent, exciting, distracting, unbearable, dividing. It is a universal language, a powerfully effective conveyor of passions. An aesthetic and social construct, music is about us, and helps us understand, or transcend the world we live in. As such, it is one of the highest forms of play. Music is also about those sounds we leave out, and as some artists have suggested, those we have no control over: the music that is produced is not the same as the music that is listened to.

A perennial and recurrent project of choice among PolyPlay students, music has often been a battleground for music neophytes trying to rest the unreachable virtuoso ghost. The music play section features musical play-things derived from simple physical principles: excitation, transmission, vibration, resonance, etc., or digital apps using the endless possibilities of smart device interactions. Cage's redefinition of modern music, Hoffnung's symphonic creature comfort antics, or art unit Maywa Denki's surrealist musical objects open designers' perspectives on new instrumental possibilities.

Taste Play

Designing for Taste play is a challenge few designers take up; and for good reasons: the category is often referred to by industry insiders as a 'liability industry'—hence this category showcases the smallest selection of playworks. The product range in this area is more often than not limited to table ware, candy products sold with separate premium items, or cooking implements—the latter requiring as little usage of electricity as possible, avoiding heating devices, and not having to involve too much time spent preparing outcomes before the actual play starts. One easy way out of such limitations has been to insist on parental supervision. The other is to design sets that do not involve electromechanical hardware. Gelatin, cold infusions, or simply make-believe, for with the latter air never tastes any better. Many projects were inspired by the Futurists' Cookbook (Marinetti 1932), and the experiments carried out by English chef Heston Blumenthal in the past decade.

Conclusion and Prognosis

Cross (1982) recognized the challenge lying in discerning 'designerly' ways of knowing from 'scientific' or 'artistic.' He suggested that since this 'third, "material" culture' concerned technology, rather than design, it was the culture of the technologist, integrating the designer, doer and maker. 'Technology involves a synthesis of knowledge and skills from both the sciences and the humanities, in the pursuit of practical tasks; it is not simply "applied science", but "the application of scientific and other organised knowledge to practical tasks".'

PolyPlay operates in a unique context, that of PolyU Design, which, while nurturing a student population made up of over 95% HK Chinese, boasts Asia's most ethnically diverse design faculty. This is an opportunity to reinvent design, playing with the metaphors of different cultural codes, a hybridization of cultural signs in a de facto cultural design lab. Play's empathetic framework enriches design, favoring multicultural perspectives, while fostering conditions necessary for the transformation of cultural identities. HK's vanguard 'Special' perspective on design, informed by Chinese heritage and free access to international worldviews, will serve as a model for emerging design practices across the region. 'PolyPlayfully way of knowing' integrating play with design, enriches research, consultancy, and pedagogy, and will serve best design practices—the simultaneous application of multiple forms of organized knowledge to practical tasks—for the invention of tomorrow's products, environments and systems.

That play and design share common features situates design in a broader conceptual context, explicating how it shapes culture, thus contributing to the study of the nature of design and facilitating design thinking education. Using the friendly Trojan hobby horse of play rationalizes design's complexities by setting a familiar philosophical ground: we connect naturally to play. Play helps assert design practice's humanistic purpose. Design is a 'conversation' between stakeholders, a semantic process shaping culture, generating meaning. Now online, anyone contributes to it, from anywhere, all the time, continuously shifting cultural paradigms within the postmodern globalized economic situation. How to explicate design's ability to navigate uncertainty, as it shapes the ambiguity of consumer society's hyperreality (Guy Debord's 'Spectacle'), an artificial reality that has gone cyber and is lived online? As play 'pretends to pretend' and tests out biological and psychological possibilities, its ambivalent yet socially appropriate framework nurtures design's ability to navigate such uncertainty. Elucidating design's diachronic (narratives) and synchronic (activities) paradigms sets the ground for a philosophical appreciation of design in the light of play, so as to enforce its role as a positive agent of socio-cultural transformation.

Designers facilitate meaningful adoption of technology: as 'play masters,' designers temper the affect of technology, designers tamper the effect of technology, harmonizing sensation (tempering affect) and thought (tampering effect) as they shape culture. Their practice is a willful participation in an enjoyable and gratifying activity; where the design project, akin to a 'magic circle,' is a temporary world within the 'real' one; where time and space is reinvented so as to focus and engage in creative flow. In this conjectural problem space, borders with 'real life' are porous, making reinterpretation and restructuration of the latter possible. Upon exit of this 'bracketed' experience, problems are governed by new rules, and purposefully reorientated towards a goal (a solution), determined by a heightened appreciation of sociocultural constraints: as one designs, one goes in and out of play to better deal with the real world. While design practice's solution focus is aimed at 'scaling up' and production, design labs are privileged environments where exploratory, critical, or speculative design, and mistakes—resonating 'in earnest' with designers' souls—should be encouraged. This nurtures iterative creation of user-empathetic 'what if' scenario development for desirable futures.

Research and validation of the utility of play in design are in their early stages. Nevertheless as Rand, Manu, Gaver, von Oech, Brown, Kristiansen, and this author have experienced, design research, practice, exploration, and education are enhanced by the mindful integration of play. Play's power to promote self-actualization enriches one's design practice. Its ontological nature transcends design's functionalism. Designing for play may not be a form of art, but it is as close as design gets to art, while remaining clearly in the realm of design. Munari (1966), who promoted design as art, points out that the ancient Japanese word for art, asobi, also means game, and this was the way he proceeded, as if playing a game, trying things out to see what would happen.

The jury is still out for a consensus on design methodology. Successful 'method' touts either post-rationalize their claim or confess their approach is not necessarily replicable by others. Play's ambiguity accommodates design's uncertainty. For instance, PolyPlay fosters conditions for:

1. Creative alternatives—in process and ideas
2. Inclusive development—where user-centeredness has no hidden agenda
3. A subversive agenda—decoupling design from consumerism.

People are not machines: they are complex adaptive systems interacting with shifting social environments. To paraphrase Manu, the ecology of play is the ecology of possibility: to the inert nature of things, play brings life. In the age of postindustrial design and the new economics, our role as design educators is no longer to facilitate the design of interactions with products or interfaces but to facilitate design interactions with possibilities. PolyPlay enforces designers' realization that their practice is a rich, purposeful, transformative, and progressive profession. To borrow from Sutton-Smith, there is no reason for an active designer to be depressed.

Acknowledgements The author wishes to thank Ilpo Koskinen and Matthew Turner for their suggestions and for advice on editing this paper.

References

Archer, B. (1979). The three Rs. *Design Studies, 1*(1), 17–20.

Baudelaire, C. (1853). Morale du joujou ('Morality of the toy'). In *Le Monde Littéraire*, 17 April 1853. Retrieved December 13, 2016, from http://www.bmlisieux.com/litterature/baudelaire/moraljou.htm.

Brown, T. (2008). Tales of creativity and play. In *2008 at Serious Play Conference*. Retrieved February 23, 2015, from http://www.ted.com/talks/tim_brown_on_creativity_and_play.

Caillois, R. (2001). *Man, play and games*. University of Illinois Press (first published 1958 in French by Gallimard as *Les jeux et les hommes*).

Carelman, J. (1969). *Catalogue d'objets introuvables*. Balland.

Chawla, L. (1998). Significant life experiences revisited: A review of research on sources of proenvironmental sensitivity. *Journal of Environmental Education, 29* (3), 11–21.

Cross, N. (1982). Designerly ways of knowing. *Design Studies, 3*(4), 221–227.

Debord, G. (1996). Exercise de la ssychogeographie. In *Internationale lettriste, potlatch 1954–1957* (pp. 11–12) (first published 1954 in French in Potlatch No. 2).

del Vecchio, G. (2003). *The blockbuster toy! How to invent the next big thing*. Pelican Publishing.

Erikson, E. H. (1993). *Childhood and society*. W. W. Norton & Company (first published 1950).

Fallman, D. (2008). The interaction design research triangle of design practice, design studies, and design exploration. *Design Issues, 24*(3), 4–18. Cambridge, MA: MIT Press.

Froebel, F. (2005). *The education of man*. Dover (first published 1826 in German as Die menschenerziehung, die erziehungs-, unterrichts- und lehrkunst, angestrebt in der allgemeinen deutschen Erziehungsanstalt zu Keilhau. Erster Band. Keilhau-Leipzig).

Froebel, F. (2015). Brief history of the kindergarten. Retrieved February 22, 2015, from http://www.froebelgifts.com/history.htm.

Gero, J. S. (1990). Design prototypes: A knowledge representation schema for design. *AI Magazine, 11*(4), 26–36.

Good, A., & Tom Tit. (2013). *Scientific amusements*. CreateSpace Independent Publishing Platform (first published 1843 in French in the journal *L'Illustration* as La science amusante).

Gupta, A. *Arvind Gupta Toys*. Retrieved December 13, 2016, from http://www.arvindguptatoys.com/.

Heskett, J. (2008). Creating economic value by design. *International Journal of Design, 3*(1), 71–84.

Huizinga, J. H. (1992). *Homo ludens: A study of the play-element in culture.* Beacon Press (first published 1938 in Dutch by Groningen, Wolters-Noordhoff cop. as *Homo ludens: proeve ener bepaling van het spelelement der cultuur*).

Kristiansen, P., & Rasmussen, R. (2014). *Building a better business using the lego serious play method.* Wiley.

Lasn, K. (Ed.). (2006). *Design anarchy.* Adbusters Media Foundation.

Leclerc, R. (2008). Character toys: Playing with identity, playing with emotions. In *Design and emotion 2008 conference proceedings.*

Leclerc, R. (2010). Hong Kong Hackshops! Creative instant toy design workshops. *International Journal of Arts and Technology, 3*(1), 63–66.

Leclerc, R. (2014). *Think, make, polyplay: A laboratory for design, play, and education.* Paper presented at the International Toy Research Association 2014 World Congress Toys as Language and Communication, Braga, PT. Retrieved from http://polyplay.hk/.

Leclerc, R., & Wan, B. (2012). Design play—Interactive education toolkit for design project planning and management. In *DesignEd Asia conference proceedings, PolyU Design.*

Manu, A. (1995). *Tooltoys: Tools with an element of play.* Copenhagen: Dansk Design Center.

Margolin, V. (2007). Design, the future and the human spirit. *Design Issues, 23*(3), 4–15 (Cambridge, MA: MIT Press).

Marinetti, F. T. (1991). The Futurist cookbook. Chronicle Books (first published 1932 in Italian by Sonzogno as La cucina Futurista).

Montessori, M. (2004). Exercises of practical life. In G. Gerald Lee (Ed.), *The Montessori method: The origins of an educational innovation: Including an abridged and annotated edition of Maria Montessori's the Montessori method.* Rowman & Littlefield. Retrieved February 22, 2015, from http://www.arvindguptatoys.com/arvindgupta/montessori-new.pdf.

Munari, B. (2008). *Design as art.* Penguin Classics (First published 1966 in Italian by Laterza as *Arte come mestiere*).

Pearce, C. (2006). Games as art: The aesthetics of play. *Visible Language*, special issue, *40*(1) (Rhode Island School of Design).

Poe, E. A. (1845). *Man of the crowd.* Retrieved November 13, 2015, from http://poestories.com/read/manofthecrowd.

Rand, P. (2014). Design and the play instinct. In *Thoughts on design*, Chronicle Books (First published 1965 by George Braziller in The education of vision, from the Vision + Value series, New York, pp. 156–174).

Rith, C., & Dubberly, H. (2007). Why Horst W. J. Rittel matters. *Design Issues, 23*(1), 72–91 (Cambridge, MA: MIT Press).

Rousseau, J. J. (1979). *Emile, or on education*. Basic Books (first published 1762 in French as *Émile ou de l'éducation*).
Salen, K., & Zimmerman, E. (2003). *Rules of play: Game design fundamentals*. Cambridge, MA: MIT Press.
Simon, H. A. (1996). *The sciences of the artificial*. Cambridge, MA: MIT Press (First published 1969).
Sutton-Smith, B. (1997). *The ambiguity of play*. Cambridge, MA: Harvard University Press.
Sutton-Smith, B. (1986). *Toys as culture*. Gardner Press.
Szent-Gyorgyi, A. (1963). Bioenergetics part 2, 57. In I. J. Good (Ed.), *The scientist speculates*. Capricorn.
Vogler, C. (1998). *The writer's journey: Mythic structure for writers*. Michael Wiese Productions.
von Oech, R. (1983). *A whack on the side of the head: How to unlock your mind for innovation*. New York: Warner Books.
von Schiller, J. C. F. (2004). Letters upon the aesthetic education of man. Letter XV (First published 1793 in German as *Brieven over de esthetische opvoeding van de mens*).
Weller, P., & The Jam. (1981). That's entertainment. In *Sound affects*. Polydor.

Author Biography

Rémi Leclerc runs PolyPlay Lab, and holds a Master's in Industrial Design, Les Ateliers—ENSCI, Paris, and is a Ph.D. candidate at Sorbonne Paris Cité University France.

Subject Index

L. Magalhães and J. Goldstein (eds.), *Toys and Communication*,
https://doi.org/10.1057/978-1-137-59136-4

Author Index

L. Magalhães and J. Goldstein (eds.), *Toys and Communication*,
https://doi.org/10.1057/978-1-137-59136-4